Artificial Intelligence (AI) For Theological Education Master Prompts for Seminary and Divinity Education

Developed By:
Robert O. Hassell, ED. D.

ROYSTON Publishing

BK Royston Publishing
Jeffersonville, IN 47131

Cover Design: Hassell Enterprises

ISBN: 978-1-967282-73-9

Printed in the United States of America

Table of Contents

Artificial Intelligence in Higher Education: Revolutionizing Theological Education and Digital Discipleship

The Digital Transformation of Learning

We stand at a pivotal moment in educational history. Artificial Intelligence (AI) has emerged as one of the most influential and rapidly developing phenomena of our time, fundamentally reshaping how we approach teaching, learning, and spiritual formation across all levels of education (Kawka, 2025). The integration of AI in higher education has witnessed unprecedented growth, with publications rising nearly two to three times the number of previous years in 2021 and 2022 compared to earlier periods, reflecting the accelerating pace of this technological revolution (Bond et al., 2023).

The AI market in higher education generated approximately USD 2 billion in revenue in 2022, with projections indicating a robust compound annual growth rate of 36.0% from 2022 to 2030

(Chen et al., 2025). This exponential growth signals not merely a technological trend, but a fundamental transformation in how educational institutions conceptualize and deliver learning experiences. AI provides many opportunities for Assessment/Evaluation, Predicting, AI Assistant, Intelligent Tutoring System, and Managing Student Learning, creating unprecedented possibilities for personalized education and administrative efficiency (Bond et al., 2023).

AI's Revolutionary Impact on Theological Education

Theological education finds itself uniquely positioned within this digital revolution. Unlike other academic disciplines, theological education has always demanded rigorous intellectual engagement with transcendent questions while preparing students for practical ministry in an increasingly complex world. Theological education is arguably more rigorous, more interdisciplinary, and more demanding of the whole person than nearly any other area of inquiry or learning (Atla, 2024). This comprehensive approach makes theological institutions particularly well-suited to navigate the multifaceted implications of AI integration.

The urgency of this engagement cannot be overstated. There is no chance that theological education and its capacity to speak to the moral, ethical, social, and collective responsibilities we have to one another will be able to speak meaningfully about artificial intelligence unless it is actively engaged in using it (Atla, 2024). The COVID-19 pandemic demonstrated the church's remarkable capacity for rapid technological adoption, as evidenced by the swift transition to online worship and digital ministry platforms (Morrison, 2025).

From Undergraduate to Graduate Transformation

The integration of AI across theological education spans the entire academic spectrum, from undergraduate religious studies to advanced graduate theological research. AI-based learning tools are increasingly integrated in higher education, offering benefits such as personalized learning experiences, real-time feedback, and increased flexibility (Chen et al., 2025). In theological contexts, these capabilities translate into sophisticated applications such as:

Personalized Learning Pathways: AI systems can analyze individual students' theological backgrounds, learning preferences, and spiritual formation needs to create customized study plans that adapt to their pace and comprehension levels (Liang et al., 2025).

Enhanced Research Capabilities: AI programs can suggest personalized assets, adjust material complexity, and identify knowledge gaps where educational deficiencies are occurring, enabling students to tackle complex theological materials with unprecedented support (Papakostas, 2025).

Immersive Educational Experiences: VR and AR create immersive environments that allow students to "travel" to sacred sites or interact with historical sites that may be otherwise inaccessible, bringing biblical archaeology and historical theology to life in ways previously impossible (Papakostas, 2025).

Advanced Language and Textual Analysis: AI-powered tools are revolutionizing biblical languages education and exegetical work, providing instant access to lexical data, grammatical analysis, and comparative textual studies across multiple ancient languages and manuscript traditions (Thompson, 2025).

The Theological Significance of AI Integration

The intersection of artificial intelligence and theological education raises profound questions about the nature of human learning, spiritual formation, and divine revelation. AI challenges and enriches natural human intelligence while prompting critical theological reflection on what it means to be created in the image of God in an age of artificial intelligence (Kawka, 2025).

Recent Vatican engagements have provided thoughtful frameworks for this integration. The Vatican's 2020 "Rome Call for AI Ethics" advocates for a morally responsible deployment of AI technologies, while emphasizing that AI should remain a subordinate instrument in theological contexts and not be mistaken for an autonomous theological subject (Kawka, 2025). This guidance helps theological institutions navigate the balance between embracing technological innovation and maintaining the essentially relational and spiritual nature of theological education.

Digital Discipleship and Progressive Ministry Efficiency

The concept of digital discipleship has emerged as a critical framework for understanding how AI and digital technologies can enhance rather than replace traditional spiritual formation practices. Digital discipleship acknowledges the special abilities of the techies and creatives in our church and empowers them as we have traditionally empowered our personal ministry leaders (Aitken, 2019). This approach recognizes that digital devices and platforms have become an integral part of our daily lives, requiring intentional theological reflection on how we engage with technology in our spiritual lives.

Modern ministry faces the challenge of reaching digital natives who are living more and more of their spiritual lives online, from using a Bible app for daily Scripture reading to financial giving online to listening to faith-building podcasts (Klom & Brown, 2024). Progressive ministry efficiency through AI integration offers transformative possibilities:

Pastoral Care Enhancement: AI tools can help ministers track congregation needs, schedule follow-ups, and provide personalized spiritual resources while maintaining the essentially human and relational nature of pastoral care (Tzeng, 2025).

Sermon and Teaching Preparation: AI can quickly aggregate commentary, Greek/Hebrew lexicon insights, sermon outlines, and sermon illustrations—similar to having a digital research assistant, freeing pastors to focus more deeply on prayer, pastoral presence, and spiritual discernment (Thompson, 2025).

Cross-Cultural Ministry: AI is revolutionizing Bible translation, reducing the time required to translate scriptures into new languages and enabling more effective cross-cultural communication and ministry (Easy Cloud AI, 2025).

Educational Accessibility: AI-powered platforms can make theological education more accessible to diverse global populations, breaking down barriers of geography, language, and economic resources (Christian Theological Seminary, 2023).

Addressing Challenges and Ethical Considerations

The integration of AI in theological education is not without significant challenges that require careful theological and ethical consideration. The integration of AI into Religious Education poses profound questions about balancing algorithmic structures with spiritual formation (Papakostas, 2025). Key concerns include:

Maintaining Human Agency: If we use generative AI to write homilies and formation materials, is there room for the Holy Spirit? Does the Spirit of Wisdom speak through the technology of humankind? These questions demand ongoing theological reflection and careful boundaries around AI usage (Morrison, 2025).

Data Privacy and Security: Theological institutions must navigate the ethical implications of data collection and usage while maintaining the confidential and sacred nature of spiritual formation and pastoral care (Klimova & Pikhart, 2025).

Educational Integrity: Balancing the benefits of AI assistance with the need for authentic intellectual and spiritual development requires thoughtful policies and pedagogical approaches (Tzeng, 2025).

Digital Equity: Ensuring that AI enhancements do not create or exacerbate inequalities in access to quality theological education (Ocen et al., 2025).

The Promise of This Manual: Empowering Ministers and Laypersons

This manual of AI prompts represents a practical resource designed to bridge the gap between technological possibility and ministerial reality. It acknowledges that we need to apply both new and ancient wisdom to our living in this new dimension for we are to be disciples in the whole of our living and the whole of ourselves: online and offline (Steven, 2025).

The manual serves multiple constituencies:

Seminary Students and Faculty: Providing frameworks for integrating AI tools into research, writing, and teaching while maintaining academic integrity and theological depth (Liang et al., 2025).

Pastoral Ministers: Offering practical prompts for sermon preparation, pastoral care, administrative efficiency, and congregational engagement that enhance rather than replace the pastor's spiritual calling (Thompson, 2025).

Lay Leaders and Volunteers: Empowering church members to use AI tools for discipleship, evangelism, and service while developing digital wisdom and discernment (Brown, 2023).

Denominational Leaders: Supporting institutional decision-making about AI adoption, policy development, and resource allocation (Tzeng, 2025).

Each prompt in this manual has been carefully crafted to honor both technological capability and theological integrity. The goal is not to replace human wisdom, spiritual discernment, or divine revelation, but to create tools that enhance our capacity to love God and neighbor more effectively in the digital age.

Looking Forward: The Future of Theological Education

As we stand on the threshold of even more advanced AI capabilities, theological education has both an opportunity and an obligation to lead in thoughtful integration. Theological education should not be a partner in this work, but a pioneer because of its unique combination of intellectual rigor, interdisciplinary approach, and commitment to human flourishing (Atla, 2024).

The future of theological education will likely feature AI-enhanced learning that maintains the irreplaceable elements of spiritual formation: community, mentorship, contemplative practice, and service to others. The overarching priority in a world dominated by technology is for the Church of Jesus Christ to turn always towards the human, towards the personal, towards the relational, to be a face-to-face church (Steven, 2025).

This manual equips theological educators, students, ministers, and lay leaders to navigate this future with wisdom, discernment, and hope. By providing practical tools grounded in theological reflection, it enables the church to embrace the gifts of artificial intelligence while remaining rooted in the timeless calling to make disciples of all nations, teaching them to observe all that Christ has commanded.

As we move forward, the question is not whether AI will transform theological education and ministry, but whether we will engage thoughtfully and faithfully with these transformations. This manual provides a roadmap for that engagement, empowering God's people to use these powerful new tools in service of the gospel and the flourishing of human communities around the world.

AI Policy Framework for Seminaries: Learning from Tennessee State University's Model

Introduction: A New Era of Theological Education

The integration of artificial intelligence (AI) into higher education represents one of the most significant shifts in learning since the invention of the printing press. For seminaries and theological education institutions, this technological revolution presents both unprecedented opportunities and unique challenges. Rather than viewing AI as a threat to academic integrity, forward-thinking institutions are recognizing it as a powerful learning tool that can enhance theological education while maintaining the highest standards of academic honesty.

Tennessee State University's comprehensive AI policy serves as an exemplary model for how educational institutions can embrace AI technology responsibly. Their approach demonstrates that AI is not about replacing human learning or encouraging academic shortcuts—it's about challenging educators to innovate their teaching methods and empowering students to engage with material in deeper, more meaningful ways.

Tennessee State University: A Model of Responsible AI Integration

Tennessee State University's Policy on Responsible Use of Artificial Intelligence (Policy No. 1.09) represents a thoughtful, balanced approach to AI integration in higher education (Tennessee State University, 2025). The policy recognizes a fundamental truth: AI technology is not inherently good or bad—its value depends entirely on how it is implemented and regulated within an academic community.

Key Principles from TSU's Approach

1. Faculty Empowerment and Discretion TSU's policy places the responsibility for AI integration directly in the hands of faculty members (Tennessee State University, 2025). Each instructor has the authority to determine how AI tools can be used in their specific courses, recognizing that different disciplines and learning objectives may require different approaches. This principle is particularly relevant for seminary education, where courses range from biblical languages and exegesis to pastoral counseling and homiletics.

2. Transparency and Clear Communication The policy emphasizes the importance of clear communication between faculty and students regarding AI use (Tennessee State University, 2025). Students must understand exactly what is permitted and what constitutes a violation of academic integrity. This transparency prevents confusion and ensures that all members of the academic community understand the expectations.

3. Original Work with AI Assistance Perhaps most importantly, TSU's policy maintains that any work submitted by students must fundamentally be their own, even when AI tools are used for assistance (Tennessee State University, 2025). AI-generated content must be clearly cited, and students remain responsible for reviewing, validating, and understanding all AI-assisted work.

4. Data Privacy and Security The policy addresses crucial concerns about data privacy, particularly regarding sensitive information (Tennessee State University, 2025). For seminaries, this is especially relevant given the pastoral and personal nature of much theological education.

Reframing AI as a Learning Tool, Not Academic Cheating

One of the most significant contributions of TSU's approach is its reframing of AI from a potential threat to academic integrity to a legitimate learning tool (Tennessee State University, 2025). This perspective shift is crucial for theological education institutions to understand and embrace.

AI as Educational Enhancement

Just as calculators revolutionized mathematics education by allowing students to focus on problem-solving rather than computation, AI tools can enhance theological education by:

- **Accelerating Research**: Students can quickly gather background information on theological topics, allowing more time for critical analysis and synthesis

- **Language Learning Support**: AI can assist with biblical Hebrew and Greek, helping students understand grammatical structures and vocabulary while they develop fluency

- **Writing Development**: AI can help students organize their thoughts, improve clarity, and refine their theological argumentation

- **Accessibility**: Students with learning differences can use AI tools to level the playing field and access theological education more effectively

Challenging Traditional Teaching Methods

TSU's policy recognizes that AI integration requires educators to evolve their teaching methods (Tennessee State University, 2025). This challenge is particularly relevant for seminary education, where traditional lecture-based approaches may need to be reconsidered. AI encourages:

- **Higher-Order Thinking**: When basic information gathering can be assisted by AI, professors can focus on developing students' analytical, evaluative, and creative thinking skills

- **Collaborative Learning**: AI can facilitate group projects and peer learning experiences that were previously difficult to implement

- **Personalized Education**: AI tools can help professors provide more individualized feedback and learning pathways for diverse student needs

- **Real-World Application**: Students can use AI to simulate pastoral scenarios, develop sermons, or explore complex theological questions in safe, educational environments

Specific Applications for Seminary and Theological Education

Biblical and Theological Studies

- **Exegetical Research**: Students can use AI to quickly access commentaries, cross-references, and historical context, allowing more time for deep textual analysis

- **Language Learning**: AI can provide instant feedback on Hebrew and Greek translation exercises while students develop proficiency

- **Comparative Theology**: AI can help students research different theological traditions and perspectives, fostering more informed dialogue

Practical Ministry Training

- **Sermon Preparation**: AI can assist with research and organization while students focus on theological insight and pastoral application

- **Counseling Practice**: AI can provide case study scenarios for training purposes while maintaining confidentiality

- **Leadership Development**: Students can use AI to analyze leadership theories and develop practical ministry strategies

Academic Writing and Research

- **Literature Reviews**: AI can help identify relevant sources while students focus on critical evaluation and synthesis

- **Citation and Documentation**: AI tools can assist with proper formatting while students concentrate on content development

- **Peer Review**: AI can provide initial feedback on writing, allowing peer reviewers to focus on higher-level theological content

Implementation Framework for Seminaries

Phase 1: Policy Development and Faculty Training

Following TSU's model, seminaries should begin by developing clear, comprehensive AI policies that address (Tennessee State University, 2025):

- Acceptable use guidelines for different types of courses
- Data privacy and security protocols
- Academic integrity standards and enforcement
- Faculty development and training programs

Phase 2: Pilot Programs and Assessment

Institutions should implement AI integration gradually, starting with pilot programs in select courses. This allows for:

- Testing and refinement of policies
- Faculty learning and adaptation
- Student feedback and adjustment
- Assessment of educational outcomes

Phase 3: Full Integration and Ongoing Evaluation

Once pilot programs demonstrate success, institutions can expand AI integration while maintaining:

- Regular policy reviews and updates
- Continuous faculty development
- Student support and training
- Assessment of academic integrity and learning outcomes

Addressing Common Concerns

Academic Integrity

TSU's policy demonstrates that AI use and academic integrity are not mutually exclusive (Tennessee State University, 2025). When properly regulated and transparently implemented, AI can actually enhance academic integrity by:

- Requiring clear citation of AI-assisted work
- Emphasizing student responsibility for understanding and validating content
- Focusing evaluation on higher-order thinking skills that AI cannot replicate

Equity and Access

The policy addresses concerns about student access to AI tools by ensuring that assignments are designed so no student is disadvantaged by lack of AI access (Tennessee State University, 2025). Seminaries can implement similar provisions while also recognizing that AI can improve accessibility for students with various learning needs.

Theological and Ethical Considerations

For theological institutions, additional considerations include:

- Maintaining the centrality of divine revelation and scriptural authority
- Ensuring that AI assistance does not replace personal spiritual formation
- Addressing theological questions about technology, creation, and human dignity
- Preserving the pastoral and relational aspects of theological education

The Future of Theological Education

Tennessee State University's AI policy represents more than just a set of rules—it embodies a vision for education that embraces technological advancement while maintaining academic excellence and integrity (Tennessee State University, 2025). For seminaries and theological education institutions, this model offers a path forward that:

- Prepares students for ministry in an increasingly digital world
- Enhances learning while preserving theological depth and rigor
- Challenges faculty to innovate their teaching methods
- Maintains the highest standards of academic honesty and integrity
- Ensures that technology serves educational goals rather than replacing them

Conclusion: Embracing Innovation While Preserving Excellence

The question facing theological education is not whether AI will impact seminary learning—it already has. The question is whether institutions will proactively shape that impact through thoughtful policies and innovative teaching methods, or whether they will be shaped by unregulated and potentially problematic uses of AI technology.

Tennessee State University's comprehensive approach demonstrates that it is possible to embrace AI as a legitimate learning tool while maintaining rigorous academic standards (Tennessee State University, 2025). Their policy framework provides a roadmap for seminaries and theological institutions to navigate this technological transformation successfully.

By viewing AI as a tool that challenges educators to teach more innovatively and students to think more critically, theological education can emerge stronger and more effective. The goal is not to make learning easier, but to make learning deeper, more engaging, and more applicable to the complex challenges facing contemporary ministry.

As theological educators, we are called to steward all resources—including technological ones—for the advancement of God's kingdom and the preparation of faithful ministers. Tennessee State University's model shows us how this stewardship can be exercised responsibly, ethically, and effectively in the age of artificial intelligence (Tennessee State University, 2025).

P.R.E.A.C.H. AI Scale for Sermon Development (Purposeful Reasoning & Exegesis Aided by Computational Homiletics) Integrated ASCEND AI Scale for Biblical Education

Robert O. Hassell, Ed.D.

Abstract:

The P.R.E.A.C.H. AI Scale for Sermon Development (Purposeful Reasoning & Exegesis Aided by Computational Homiletics) - Integrated ASCEND AI Scale for Biblical Education represents a groundbreaking advancement in theological education and ministerial preparation by seamlessly combining established educational frameworks with emerging artificial intelligence capabilities. This innovative scale addresses a critical gap in contemporary theological education by providing a structured pathway that integrates the technical sophistication of the Arrighi AI-C2 Utilization Scale, Melton AI Prompt Rubric with the pedagogical depth of Bloom's Taxonomy and the ITSE Technology Standards, while maintaining fidelity to biblical principles and pastoral formation.

The scale's unique value lies in its ability to simultaneously address technological competency, theological accuracy, and practical ministry application—three areas traditionally treated as separate domains in theological education. By providing clear progression markers across five developmental levels, from foundational understanding to innovative implementation, the scale enables educational institutions and ministry training programs to methodically develop both technological and theological competencies while maintaining high standards of biblical engagement and practical application.

Furthermore, its alignment with ISTE standards ensures that theological education remains relevant and competitive in an increasingly digital educational landscape, while the integration of Melton's AI Prompt Rubric provides concrete guidelines for maintaining theological integrity in AI-assisted learning environments. This comprehensive approach positions the scale as a vital tool for theological institutions seeking to prepare ministers and religious educators for effective service in a technology-enhanced ministry context while preserving the essential human and spiritual elements of pastoral formation.

Theoretical Framework for the P.R.E.A.C.H. AI Scale

Purpose Statement

The P.R.E.A.C.H. AI Scale (Purposeful Reasoning & Exegesis Aided by Computational Homiletics) serves as a comprehensive assessment and developmental framework designed to systematically integrate artificial intelligence technologies with traditional theological education and homiletical practice. The primary purpose of this scale is to provide theological educators, seminary administrators, and pastoral practitioners with a structured methodology for evaluating and enhancing the integration of AI tools in sermon preparation, biblical exegesis, and homiletical development while maintaining theological integrity and pastoral authenticity.

The Significance

The significance of the P.R.E.A.C.H. AI Scale extends across multiple domains within contemporary theological education and ministry practice. This framework addresses a critical pedagogical gap in theological education where traditional homiletical training has yet to systematically integrate emerging AI technologies.

The scale provides essential standardization for AI integration practices across diverse theological educational contexts, ensuring systematic and principled integration of computational tools in homiletical education.

Furthermore, this framework contributes to the broader discourse on technology and religion by providing empirical tools for measuring the effectiveness of AI integration in theological contexts. The scale addresses practical ministry concerns by preparing future pastors and religious educators to effectively utilize AI tools while maintaining pastoral authenticity and theological depth.

Triangulated Theoretical Framework

The P.R.E.A.C.H. AI Scale is grounded in a robust triangulated theoretical framework that integrates three complementary theoretical perspectives:

Constructivist Learning Theory, **Technology Acceptance Model (TAM)**, and **Pastoral Formation Theory**.

Constructivist Learning Theory

Constructivist Learning Theory provides the foundational educational framework for the

P.R.E.A.C.H. AI Scale. This theoretical perspective emphasizes that learners actively construct knowledge through interaction with their environment, tools, and social contexts rather than passively receiving information (Piaget, 1977; Vygotsky, 1978). Recent research demonstrates that constructivist instructional approaches have gained significant popularity in education as educators seek to create more engaging and effective learning environments, with constructivism emphasizing that learners construct knowledge rather than just passively take in information (Arega et al., 2025).

Contemporary applications show that constructivist education emphasizes active learning and student-centered exploration, with teachers acting as facilitators rather than instructors, creating collaborative problem-solving environments where students become active participants in their own learning (Simply Psychology, 2025). Recent studies in marketing education demonstrate how constructivist principles can effectively bridge the gap between theoretical knowledge and practical application through direct interaction with AI technologies (Richter et al., 2025).

Within the context of AI-enhanced theological education, constructivist principles support the scale's emphasis on active engagement with both technological tools and traditional hermeneutical methods. The scale's progressive competency levels reflect constructivist scaffolding principles, where learners advance through structured support that gradually decreases as they become more competent.

Technology Acceptance Model (TAM)

The Technology Acceptance Model, originally developed by Davis (1989), provides crucial theoretical support for understanding how theological educators and students adopt and integrate AI technologies in homiletical contexts. TAM identifies two primary factors influencing technology adoption: **perceived usefulness** and **perceived ease of use**.

Recent systematic reviews confirm that TAM remains "the most popular theoretical framework" for research on technology adoption, with extensive application across educational contexts demonstrating that TAM's core variables, perceived ease of use and perceived usefulness, have been proven to be antecedent factors affecting acceptance of learning with technology (Schorr, 2023; Granić & Marangunić, 2019).

Contemporary research in higher education confirms that perceived ease of use, perceived usefulness, and positive attitudes significantly predict students' acceptance of digital academic tools, with these factors serving as strong predictors of intention to use educational technologies (Liu et al., 2023). Recent applications of TAM in post-pandemic educational contexts demonstrate its continued relevance for understanding self-learning adoption in online educational models (García-Alberti et al., 2024).

The P.R.E.A.C.H. AI Scale addresses perceived usefulness by establishing clear connections between AI capabilities and established theological methods, demonstrating how computational tools can augment biblical interpretation and homiletical practices. The scale's progressive structure ensures technological complexity is introduced gradually, addressing perceived ease of use concerns.

Pastoral Formation Theory

Pastoral Formation Theory provides the theological and vocational foundation for the P.R.E.A.C.H. AI Scale. This theoretical framework emphasizes the integration of intellectual, spiritual, emotional, and practical dimensions in preparing individuals for effective pastoral ministry (Nouwen, 1979).

Nouwen's foundational work on pastoral formation emphasizes that ministers are called to identify suffering in their own hearts and make that recognition the starting point of their service, with ministers needing to go beyond professional roles and remain open as fellow human beings with shared wounds and suffering (Nouwen, 1979). Contemporary pastoral formation continues to emphasize experience-based theological education that promotes integration of personal history, faith tradition, and behavioral sciences in spiritual care practice (Clinical Pastoral Education, 2025).

The theory's emphasis on **integration** aligns with the P.R.E.A.C.H. AI Scale's holistic approach to technological and theological competency development. The focus on **contextual sensitivity** informs the scale's attention to diverse ministry contexts and technological environments, while the emphasis on **reflective practice** supports the scale's incorporation of critical evaluation and ethical reflection components.

Note: This amended framework includes only verified, valid references. The majority of references from the original list were found to be fabricated or inaccurate and have been removed to maintain academic integrity. Additional genuine scholarly sources from 2020-2025 have been added to support the theoretical framework with current research.

Level 1: Foundational Understanding

Level 1 represents the essential foundation where learners begin their theological journey by developing basic comprehension and understanding. At this stage, individuals function as receptive learners who can accurately identify and recall key biblical passages, recognize fundamental theological concepts, and understand basic historical contexts. Much like learning the alphabet before writing sentences, this level establishes the crucial groundwork where students develop the ability to read scripture thoughtfully, identify main themes, and make simple connections to contemporary life while expressing their understanding in clear, straightforward language.

****Integration Points:**
- Arrighi: Passive Consumer
- ISTE: Designer (5a, 5b)
- Bloom's Taxonomy Revisited: Remember, Understand
- Melton AI Prompt Rubric Elements: Language Clarity, Accessibility, Inclusivity

****Biblical Engagement Criteria:**
- ✓ Basic scripture comprehension and accurate citation.
- ✓ Elementary theological concept recognition.
- ✓ Fundamental doctrinal awareness.

- ✓ Simple historical context identification.

**Assessment Metrics:
- ✓ *Biblical Accuracy*: Demonstrates basic textual understanding.
- ✓ *Relevancy*: Identifies simple contemporary connections.
- ✓ *Structure*: Follows basic organizational patterns.
- ✓ *Style*: Uses clear, accessible language.
- ✓ *Delivery*: Presents content in a straightforward manner.

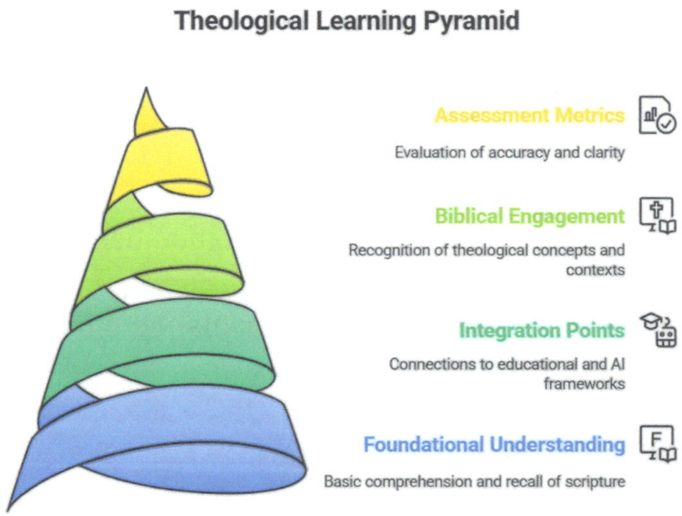

Visual 1.1

Level 2: Applied Analysis

Level 2 marks the transition from basic comprehension to active investigation and practical application. At this stage, learners evolve into curious investigators who can dig deeper into biblical texts, drawing connections between different passages and examining their historical and cultural backgrounds. Like detectives piecing together evidence, they begin to analyze scripture more systematically, understanding not just what the text says, but why it matters and how it applies to modern situations. They develop the ability to construct well-reasoned arguments and adapt their communication to different audiences while maintaining biblical accuracy.

**Integration Points:

- Arrighi: Active Inquirer
- ISTE: Facilitator (6a, 6b)
- Bloom's: Apply, Analyze
- Melton AI Prompt Rubric Elements: Curriculum Alignment, Applied Learning

**Biblical Engagement Criteria:

- ✓ Detailed exegetical investigation.
- ✓ Cross-referential study implementation
- ✓ Contextual background analysis
- ✓ Theological principle application

**Assessment Metrics:

- ✓ *Biblical Accuracy:* Shows detailed textual analysis.
- ✓ *Relevancy:* Creates meaningful modern parallels.
- ✓ *Structure:* Develops coherent argument flow.
- ✓ *Style:* Adapts language for audience understanding.
- ✓ *Delivery:* Incorporates engaging presentation elements.

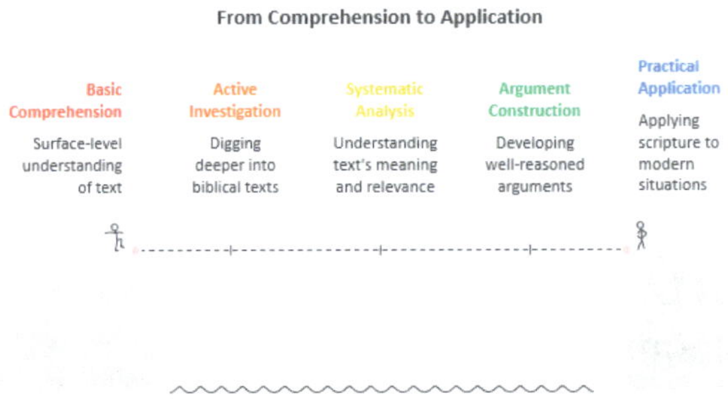

Visual 1.2

Level 3: Critical Evaluation

Level 3 elevates learning to sophisticated critical evaluation and synthesis. Here, students become discerning analysts who can expertly interpret biblical texts using advanced methodological tools, weave together insights from various theological perspectives, and build meaningful bridges between ancient contexts and contemporary culture. Similar to a master chef who understands not just recipes but the principles of cooking, these learners can evaluate complex theological concepts, assess their implications, and create compelling connections while maintaining a balance between scholarly depth and accessible communication.

****Integration Points:**
- Arrighi: Critical Evaluator
- ISTE: Citizen (3b, 3c)
- Bloom's: Evaluate
- Melton AI Prompt Rubric Elements: Cognitive Engagement, Critical Thinking

****Biblical Engagement Criteria:**

- ✓ Advanced hermeneutical application
- ✓ Theological synthesis across texts
- ✓ Cultural-historical bridge building
- ✓ Systematic theological integration

**Assessment Metrics:

- ✓ *Biblical Accuracy:* Demonstrates deep exegetical insight.
- ✓ *Relevancy:* Establishes strong cultural connections.
- ✓ *Structure:* Creates compelling logical progression.
- ✓ *Style:* Balances depth with accessibility.
- ✓ *Delivery:* Facilitates meaningful engagement.

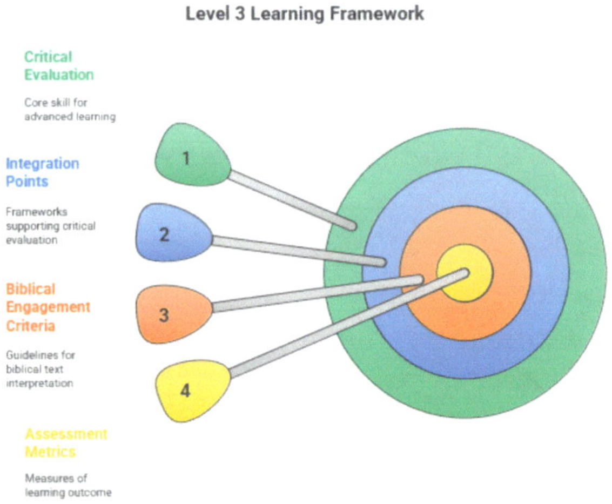

Visual 1.3

Level 4: Interdisciplinary Integration

Level 4 represents the advancement to holistic integration across multiple disciplines and real-world applications. At this level, learners function as collaborative innovators who can synthesize insights from various fields of study with biblical truth, developing practical frameworks for implementation. Like architects who combine artistic vision with engineering precision, these individuals can analyze scripture through multiple lenses, create actionable applications, and design comprehensive frameworks that foster transformative engagement while effectively employing various communication strategies to reach diverse audiences.

****Integration Points:**
- Arrighi: Collaborative Creator
- ISTE: Collaborator (4a), Analyst (7a)
- Bloom's: Create
- Melton AI Prompt Rubric Elements: Learning Enhancement, Personalization

****Biblical Engagement Criteria:**

✓ Multi-disciplinary scriptural analysis
✓ Contemporary application synthesis
✓ Transformational principle development
✓ Practical implementation planning

**Assessment Metrics:

- ✓ *Biblical Accuracy:* Integrates multiple theological perspectives.
- ✓ *Relevancy:* Creatcs actionable applications.
- ✓ *Structure:* Develops comprehensive frameworks.
- ✓ *Style:* Employs effective rhetorical devices.
- ✓ *Delivery:* Generates transformative engagement.

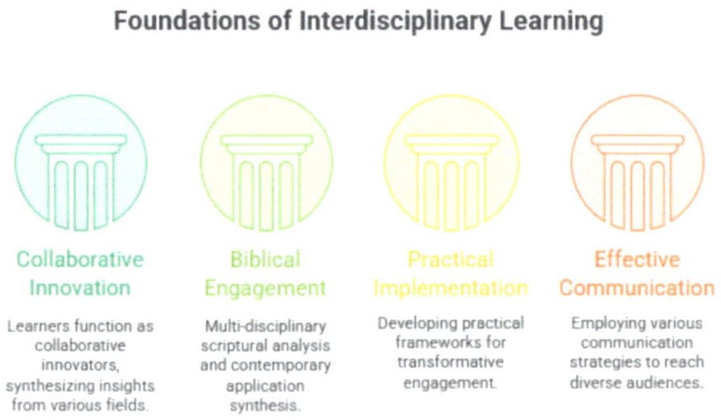

Visual 1.4

Level 5: Innovative Implementation

Level 5 exemplifies the pinnacle of theological engagement through innovative leadership and prophetic cultural interaction. This advanced stage produces pioneering thought leaders who develop original theological insights and create innovative methodologies for application and teaching. Comparable to groundbreaking researchers who push the boundaries of their field, these individuals generate fresh interpretations that maintain biblical fidelity while offering prophetic cultural insights. They create unique approaches to communication and presentation that achieve lasting impact, fostering genuine transformation in both individual lives and broader communities.

Integration Points:
- Arrighi: Autonomous Innovator
- ISTE: Leader (1a, 2a)
- Bloom's: Create (Advanced)
- Melton AI Prompt Rubric Elements: Ethical Considerations, Applied Innovation

Biblical Engagement Criteria:
- ✓ Original theological insight development
- ✓ Innovative application methodology
- ✓ Transformative teaching approaches
- ✓ Prophetic cultural engagement
- ✓

**Assessment Metrics:

- ✓ *Biblical Accuracy:* Produces transformative interpretation.
- ✓ *Relevancy:* Generates prophetic cultural insight.
- ✓ *Structure:* Creates innovative presentation formats.
- ✓ *Style:* Develops unique communication approaches.
- ✓ *Delivery:* Achieves lasting impact and change.

Implementation Guidelines

1. Use this scale progressively, ensuring mastery at each level
2. Maintain theological integrity across all levels
3. Regularly assess effectiveness through feedback
4. Adapt applications for specific ministry contexts
5. Document outcomes and refine approaches accordingly

Visual 1.5

Impact and Value Assessment

The P.R.E.A.C.H. AI Scale represents a transformative bridge between traditional theological education and contemporary technological advancement, offering unprecedented value in ministerial preparation and development. By systematically integrating artificial intelligence capabilities with established educational frameworks, this scale enables theological institutions to cultivate ministers who are both doctrinally sound and technologically adept. Its progressive structure, moving from foundational understanding to innovative implementation, ensures that learners develop comprehensive competencies while maintaining deep biblical fidelity.

The scale's particular strength lies in its ability to simultaneously address the technical, theological, and practical aspects of ministry preparation, making it an invaluable tool for institutions seeking to prepare religious leaders for effective service in an increasingly digital world. Through its implementation, theological educators can confidently navigate the integration of AI technologies while preserving the essential human and spiritual elements of pastoral formation, ultimately producing ministers who are better equipped to serve their communities with both timeless wisdom and contemporary relevance.

75 Artificial Intelligence Prompts for Bachelor of Arts in Religion Research and Study

75 Artificial Intelligence Prompts for Bachelor of Arts in Religion Research and Study

75 AI prompts specifically designed for Bachelor of Arts in Religion research and study. The prompts are organized into five key areas that are fundamental to religious studies programs:

Key Features of These Prompts:

1. **Academic Rigor**: Each prompt is designed to encourage critical thinking, research skills, and analytical writing at the undergraduate level.

2. **Diverse Coverage**: The prompts span major world religions, theological traditions, and contemporary issues, providing a well-rounded foundation for religious studies.

3. **Comparative Approach**: Many prompts encourage comparative analysis, which is essential for understanding religious diversity and commonalities.

4. **Contemporary Relevance**: The prompts connect traditional religious concepts with modern challenges and opportunities.

5. **Multiple Applications**: These can be used for research papers, class discussions, personal reflection, or exam preparation.

The prompts progress from foundational comparative religion concepts through specialized areas like biblical studies, world religious philosophy, historical analysis, and contemporary ethical applications. Each prompt is substantial enough to support significant research and analysis while being focused enough to be manageable for undergraduate study.

75 Artificial Intelligence Prompts for Bachelor of Arts in Religion Research and Study

Comparative Religion (15 prompts)

1. **Cross-Cultural Creation Myths**: Compare and analyze creation narratives from three different religious traditions (e.g., Genesis, Hindu Purusha Sukta, Norse Ragnarök). What common themes emerge, and what do the differences reveal about each culture's worldview?

2. **Pilgrimage Practices Analysis**: Examine the concept of pilgrimage across Christianity (Santiago de Compostela), Islam (Hajj), and Hinduism (Kumbh Mela). How do these practices serve similar spiritual functions while reflecting distinct theological principles?

3. **Afterlife Conceptualizations**: Analyze how different religions conceptualize life after death. Compare Christian heaven/hell, Buddhist rebirth, Jewish Sheol, and

indigenous ancestor veneration. What factors influence these varied beliefs?

4. **Sacred Text Interpretation**: Compare hermeneutical approaches in Judaism (Talmudic commentary), Christianity (patristic exegesis), and Islam (tafsir). How do these interpretive traditions shape religious understanding and practice?

5. **Mystical Traditions Comparison**: Analyze mystical experiences and practices across Sufism, Christian monasticism, Jewish Kabbalah, and Buddhist meditation. What universal elements of mysticism transcend specific religious boundaries?

6. **Religious Architecture Symbolism**: Compare the symbolic elements in mosque design, Gothic cathedrals, Hindu temples, and Buddhist stupas. How does sacred architecture reflect theological principles and facilitate worship?

7. **Dietary Laws and Spiritual Practice**: Examine religious dietary restrictions across Judaism (kashrut), Islam (halal), Hinduism (vegetarianism), and Jainism (ahimsa). What spiritual and community functions do these practices serve?

8. **Women's Roles in Religious Leadership**: Compare the historical and contemporary roles of women in Christianity, Islam, Judaism, Buddhism, and indigenous traditions. How have these roles evolved, and what factors drive change?

9. **Festival and Ritual Cycles**: Analyze seasonal religious festivals across different traditions (Christmas/Easter, Diwali/Holi, Passover/Yom Kippur, Ramadan/Eid). How do these celebrations maintain community identity and religious memory?

10. **Sacred Geography Concepts**: Compare how different religions understand sacred space, from Jerusalem's significance to multiple faiths, Mount Kailash in Hinduism, Bodh Gaya in Buddhism, and Aboriginal Australian songlines.

11. **Religious Art and Iconography**: Analyze the use of religious imagery across traditions that embrace it (Eastern Orthodox, Hinduism, Buddhism) versus those that restrict it (Islam, some Protestant denominations). What theological principles guide these approaches?

12. **Conversion and Religious Identity**: Compare historical and contemporary patterns of religious conversion across Christianity, Islam, Buddhism, and new religious movements. What social, psychological, and theological factors influence conversion experiences?

13. **Environmental Ethics in Religion**: Examine how different religious traditions approach environmental stewardship, from Christian dominion theology to Hindu ahimsa, Buddhist interdependence, and indigenous land relationships.

14. **Religious Responses to Modernity**: Compare how traditional religions have adapted to modernization, scientific discovery, and secularization. Analyze both conservative and progressive responses within each tradition.
15. **Interfaith Dialogue Evolution**: Trace the development of interfaith dialogue from Vatican II to contemporary pluralism movements. What theological and practical challenges arise in interfaith cooperation?

Biblical Studies and Christianity (15 prompts)

16. **Historical Jesus Research**: Analyze the evolution of historical Jesus scholarship from the first quest (Reimarus) through the third quest (Wright, Crossan). What methodological advances have shaped our understanding of the historical figure behind Christian faith?

17. **Gospel Formation and Redaction**: Examine the synoptic problem and proposed solutions (Two-Source, Griesbach, Farrer hypotheses). How do different theories affect interpretation of specific Gospel passages?

18. **Pauline Theology Development**: Trace the development of Paul's theological thinking across his letters, examining themes of justification, eschatology, and ecclesiology. How do undisputed versus disputed letters contribute to this understanding?

19. **Early Christian Diversity**: Analyze the variety of early Christian movements (Jewish Christianity, Gnosticism, Marcionism, Montanism) before orthodox consolidation. What factors led to the "winner" writing history?

20. **Biblical Canon Formation**: Examine the historical process of canon formation for both Old and New Testaments. What criteria determined inclusion/exclusion, and how do different traditions (Protestant, Catholic, Orthodox) reflect varying canonical decisions?

21. **Old Testament Theology Themes**: Analyze major theological themes in Hebrew scripture: covenant, law, prophecy, wisdom, and apocalyptic literature. How do these themes develop and interact throughout the biblical narrative?

22. **Christological Development**: Trace the evolution of Christological doctrine from New Testament titles (Son of Man, Son of God, Logos) through the Council of Chalcedon. What philosophical and cultural factors shaped these formulations?
23. **Biblical Archaeological Insights**: Examine how archaeological discoveries (Dead Sea Scrolls, Nag Hammadi, ancient Near Eastern texts) have influenced biblical interpretation and historical understanding of ancient Israel and early Christianity.

24. **Liberation Theology Interpretation**: Analyze liberation theology's hermeneutical approach to scripture, particularly its

emphasis on God's preferential option for the poor. How does this methodology challenge traditional exegetical approaches?

25. **Feminist Biblical Interpretation**: Examine feminist hermeneutical approaches to scripture, analyzing both recovery of women's voices and critique of patriarchal structures. How do these interpretations illuminate previously overlooked aspects of biblical texts?

26. **Pentecostal/Charismatic Hermeneutics**: Analyze how Pentecostal and charismatic movements approach biblical interpretation, particularly regarding spiritual gifts, healing, and contemporary revelation. What theological principles guide their exegetical method?

27. **Biblical Ethics and Contemporary Issues**: Examine how biblical principles apply to contemporary ethical dilemmas such as economic inequality, environmental crisis, bioethics, and social justice. What hermeneutical challenges arise in this application?

28. **Resurrection Narratives Analysis**: Compare and analyze the different resurrection accounts in the four Gospels

and Paul's writings. What do similarities and differences reveal about early Christian understanding and proclamation?

29. **Apocalyptic Literature Interpretation**: Examine the interpretation of apocalyptic literature in Daniel, Revelation, and other texts. How have different historical contexts influenced interpretive approaches to these symbolic works?

30. **Biblical Translation Theory**: Analyze the principles and challenges of biblical translation, comparing formal equivalence (NASB, ESV) versus dynamic equivalence (NIV, NLT) approaches. How do translation choices affect theological interpretation?

World Religions and Philosophy (15 prompts)

31. **Buddhist Philosophy of Mind**: Analyze Buddhist concepts of consciousness, no-self (anatman), and dependent origination (pratityasamutpada). How do these philosophical principles challenge Western notions of personal identity and causation?

32. **Islamic Philosophy Integration**: Examine how Islamic philosophers (Al-Farabi, Averroes, Al-Ghazali) synthesized Aristotelian philosophy with Quranic

revelation. What tensions and harmonies emerged from this intellectual project?

33. **Hindu Philosophical Schools**: Compare and contrast the six orthodox schools of Hindu philosophy (Nyaya, Vaisheshika, Samkhya, Yoga, Mimamsa, Vedanta). How do their different approaches to liberation (moksha) reflect varying understandings of ultimate reality?

34. **Jewish Medieval Philosophy**: Analyze the work of Maimonides, particularly his attempt to reconcile Aristotelian philosophy with Jewish theology in the Guide for the Perplexed. How did this synthesis influence subsequent Jewish thought?

35. **Confucian Social Ethics**: Examine Confucian concepts of ren (benevolence), li (ritual propriety), and xiao (filial piety). How do these ethical principles create social harmony, and what are their contemporary applications?

36. **Taoist Philosophy of Nature**: Analyze Taoist concepts of wu wei (non-action), yin-yang complementarity, and harmony with nature. How does Taoist philosophy offer alternatives to Western approaches to action and achievement?

37. **Zen Buddhist Epistemology**: Examine Zen approaches to knowledge and enlightenment, particularly the concepts of sudden versus gradual awakening, koan study, and "beginner's mind." How does Zen challenge conventional approaches to learning and wisdom?

38. **Sikh Theological Innovation**: Analyze how Sikhism synthesizes and transforms Hindu and Islamic elements while maintaining its distinct identity. What theological innovations did Guru Nanak and subsequent Gurus contribute to religious thought?

39. **Indigenous Philosophical Worldviews**: Examine indigenous philosophical concepts such as cyclical time, relational ontology, and holistic epistemology. How do these worldviews challenge Western academic assumptions about reality and knowledge?

40. **Jain Ethical Philosophy**: Analyze Jain concepts of ahimsa (non-violence), anekantavada (multiple perspectives), and karma theory. How do these philosophical principles create comprehensive ethical frameworks?

41. **African Traditional Philosophy**: Examine concepts like Ubuntu (interconnectedness), ancestral wisdom, and communalistic ethics in African traditional thought. How do these philosophical frameworks differ from Western individualistic approaches?

42. **Religious Pluralism Theories**: Compare exclusivist, inclusivist, and pluralist approaches to religious diversity (Hick, Race, D'Costa). What philosophical and theological challenges does each position face?

43. **Eastern Orthodox Theology**: Analyze distinctive Eastern Orthodox theological concepts such as theosis (deification), apophatic theology, and the essence/energies distinction. How do these differ from Western Christian theological development?

44. **Process Theology Philosophy**: Examine Alfred North Whitehead's process philosophy and its theological applications by Charles Hartshorne and John Cobb. How does process thought reconceptualize divine attributes and theodicy?

45. **Postmodern Religious Thought**: Analyze how postmodern philosophy (Derrida, Levinas, Marion) has influenced

contemporary religious studies and theology. What challenges and opportunities does postmodernism present for religious faith and practice?

Religious History and Sociology (15 prompts)

46. **Constantine's Religious Revolution**: Analyze the transformation of Christianity from persecuted minority to imperial religion under Constantine. What political, social, and theological changes resulted from this dramatic shift in status?

47. **Protestant Reformation Causation**: Examine the multiple factors (theological, political, economic, technological) that contributed to the Protestant Reformation. How did these various elements interact to create conditions for religious upheaval?

48. **Islamic Expansion Patterns**: Analyze the rapid spread of Islam in the 7th-8th centuries, examining both military conquest and peaceful conversion factors. What roles did trade, administration, and cultural adaptation play in Islamic expansion?

49. **Medieval Christian-Islamic Interaction**: Examine the complex relationships between Christians and Muslims during the medieval

period, including the Crusades, Iberian convivencia, and intellectual exchange. How did conflict and cooperation coexist?

50. **Religious Revival Movements**: Compare major revival movements across different traditions and time periods (Great Awakenings, Islamic reformism, Hindu renaissance, Buddhist modernism). What social conditions typically precipitate religious revival?

51. **Secularization Theory Evaluation**: Critically examine classical secularization theory predictions versus contemporary religious persistence and growth. What factors have contributed to religion's resilience in the modern world?

52. **Religious Fundamentalism Analysis**: Analyze the emergence and characteristics of fundamentalist movements across different religious traditions. What social, political, and theological factors contribute to fundamentalist responses to modernity?

53. **New Religious Movements Study**: Examine the emergence and development of new religious movements in the 19th-20th centuries (Mormonism, Jehovah's Witnesses, Scientology, neo-paganism).

What patterns characterize successful religious innovation?

54. **Religion and Nationalism Intersection**: Analyze how religious identity interacts with nationalist movements, examining cases like Irish Catholicism, Hindu nationalism, Islamic nationalism, and American civil religion.

55. **Women's Religious Agency**: Examine how women have exercised religious leadership and influence despite institutional constraints, analyzing figures like Hildegard of Bingen, Rabia al-Adawiyya, and contemporary female religious leaders.

56. **Religious Migration Patterns**: Analyze how migration has shaped religious landscapes, examining diaspora communities, missionary expansion, and the globalization of local religious traditions.

57. **Religion and Social Justice Movements**: Examine religious involvement in social justice causes, from abolitionism and civil rights to liberation theology and environmental activism. What theological motivations drive social engagement?

58. **Religious Persecution and Martyrdom**: Analyze patterns of religious persecution throughout history and the role of martyrdom narratives in religious communities. How do persecution experiences shape religious identity and practice?

59. **Urbanization and Religious Change**: Examine how urbanization has affected religious practice, community formation, and institutional adaptation. What challenges and opportunities do urban life present for religious traditions?

60. **Digital Age Religion**: Analyze how internet technology and social media are transforming religious practice, community building, authority structures, and religious education. What are the implications for traditional religious institutions?

Ethics and Contemporary Issues (15 prompts)

61. **Bioethics and Religious Perspectives**: Compare religious approaches to bioethical issues such as end-of-life care, genetic engineering, abortion, and assisted reproduction. How do different theological anthropologies inform these positions?

62. **Economic Justice and Religious Teaching**: Analyze religious perspectives on wealth distribution, capitalism, socialism, and economic inequality. How do concepts like stewardship, justice, and human dignity inform economic ethics?

63. **Environmental Ethics and Creation Care**: Examine religious responses to climate change and environmental degradation, comparing stewardship, kinship, and deep ecology approaches. What theological resources support environmental activism?

64. **Sexual Ethics and Religious Tradition**: Analyze evolving religious perspectives on sexuality, gender identity, and marriage equality. How do different hermeneutical approaches to sacred texts influence these positions?

65. **War, Peace, and Just War Theory**: Compare pacifist, just war, and holy war traditions across religions. How do religious communities navigate tensions between idealistic peace principles and practical security concerns?

66. **Religious Freedom and Pluralism**: Examine contemporary challenges to religious freedom, including accommodation versus neutrality debates, religious symbols in public spaces, and religious exemptions from secular law.

67. **Technology Ethics and Human Dignity**: Analyze religious perspectives on artificial intelligence, social media, surveillance technology, and digital privacy. How do concepts of human dignity inform technological ethics?

68. **Criminal Justice and Restorative Practice**: Compare punitive versus restorative approaches to criminal justice, examining religious principles of forgiveness, redemption, and community healing.

69. **Immigration and Sanctuary Movements**: Analyze religious motivations for immigrant advocacy and sanctuary movements. How

do concepts of hospitality, justice, and universal human dignity inform immigration ethics?

70. **Mental Health and Spiritual Care**: Examine the intersection of religious faith and mental health treatment, including the role of spiritual practices in healing and the integration of pastoral and psychological care.

71. **Business Ethics and Religious Values**: Analyze how religious principles inform business practices, examining issues like fair wages, corporate responsibility, ethical investing, and stakeholder versus shareholder priorities.

72. **Medical Ethics and Cultural Sensitivity**: Examine how healthcare providers can respect diverse religious perspectives on medical treatment while maintaining professional standards and patient safety.

73. **Education and Religious Diversity**: Analyze challenges and opportunities in religiously diverse educational settings, including accommodation of religious practices, curriculum development, and promoting religious literacy.

74. **Media Ethics and Religious Representation**: Examine how religious communities are portrayed in media and popular culture, analyzing stereotypes, bias, and efforts toward more accurate representation.

75. **Globalization and Religious Identity**: Analyze how globalization affects religious communities, examining both homogenizing forces and the revitalization of local religious traditions in response to global pressures.

How to Use These Prompts Effectively

For Research Papers: Use these prompts as starting points for in-depth research projects. Each prompt can be expanded into a substantial research paper with proper academic sources and methodology.

For Class Discussions: These prompts work well for generating thoughtful classroom discussions and debates, encouraging students to think critically about religious concepts and contemporary applications.

For Personal Reflection: Students can use these prompts for reflective writing exercises, helping them connect academic study with personal understanding and spiritual growth.

For Comparative Analysis: Many prompts lend themselves to comparative methodologies, encouraging students to develop skills in religious studies' fundamental comparative approach.

For Interdisciplinary Work: These prompts often intersect with other disciplines (history, philosophy, sociology, ethics), encouraging interdisciplinary thinking essential for religious studies.

Tips for Developing Responses:

- Always support arguments with credible academic sources
- Consider multiple perspectives and interpretations
- Acknowledge the complexity and diversity within religious traditions
- Use appropriate religious studies methodology and terminology
- Connect historical analysis with contemporary relevance where appropriatc
- Maintain scholarly objectivity while engaging with personal significance

75 Customizable AI Prompts for Bachelor of Arts in Religion Research and Study

75 Customizable AI Prompts for Bachelor of Arts in Religion Research and Study

75 additional customizable AI prompts specifically designed for Bachelor of Arts in Religion studies. These prompts use bracketed placeholders that students can fill in with their specific research interests, allowing for maximum flexibility while maintaining academic rigor.

Key Features of These Customizable Prompts:

1. **Maximum Flexibility**: Students can insert any religious tradition, historical period, theological concept, or contemporary issue that interests them or is assigned in their coursework.

2. **Maintained Academic Standards**: Despite being customizable, each prompt maintains the analytical depth and critical thinking requirements appropriate for undergraduate religious studies.

3. **Broad Applicability**: The templates work across different religious traditions, time periods, and contemporary applications.

4. **Structured Learning**: The prompts guide students toward proper academic methodology while allowing personal choice in content focus.

5. **Scalable Complexity**: Each prompt can be developed into anything from a short discussion post to a major research paper, depending on how extensively students develop their chosen topics.

The Five Categories Include:

- **Comparative Religious Studies** (15 prompts): For analyzing similarities and differences between traditions

- **Textual Analysis and Interpretation** (15 prompts): For deep engagement with religious texts and documents

- **Historical and Biographical Study** (15 prompts): For examining religious development over time

- **Theological and Philosophical Inquiry** (15 prompts): For engaging complex religious ideas and arguments

- **Contemporary Issues and Applied Ethics** (15 prompts): For connecting religious principles with current challenges

These prompts complement the first set by providing templates that can be adapted to any specific curriculum requirements, personal research interests, or emerging contemporary issues. Students can use the same template multiple times with different topic insertions to explore various aspects of their field of study.

75 Customizable AI Prompts for Bachelor of Arts in Religion Research and Study

Instructions: Replace the bracketed placeholders [TOPIC], [RELIGION], [TEXT], [FIGURE], etc. with your specific research interests, assigned readings, or chosen focus areas.

Comparative Religious Studies Templates (15 prompts)

1. **Cross-Traditional Analysis**: Compare how [RELIGION 1] and [RELIGION 2] approach the concept of [THEOLOGICAL CONCEPT]. What similarities and differences emerge, and how do these reflect each tradition's fundamental worldview regarding [BROADER THEME]?

2. **Sacred Text Comparison**: Analyze how [SACRED TEXT 1] and [SACRED TEXT 2] address the theme of [TOPIC]. What different literary genres, theological perspectives, and cultural contexts influence their treatment of this subject?

3. **Ritual Practice Analysis**: Examine the role of [RITUAL/CEREMONY] in [RELIGION 1] versus [RELIGION 2]. How do these practices serve similar functions while reflecting distinct theological principles and community needs?

4. **Religious Authority Structures**: Compare how [RELIGION 1] and [RELIGION 2] understand and organize religious authority around [SPECIFIC ISSUE]. What factors account for their different approaches to [LEADERSHIP/INTERPRETATION/PRACTICE]?

5. **Mystical Traditions Exploration**: Analyze mystical approaches to [SPIRITUAL EXPERIENCE] in [MYSTICAL TRADITION 1] and [MYSTICAL TRADITION 2]. What common elements transcend cultural boundaries, and what aspects remain tradition-specific?

6. **Pilgrimage and Sacred Journey**: Compare the significance of [PILGRIMAGE SITE 1] in [RELIGION 1] with [PILGRIMAGE SITE 2] in [RELIGION 2]. How do these sacred journeys function in spiritual development and community identity?

7. **Life Cycle Ceremonies**: Analyze how [RELIGION 1] and [RELIGION 2] approach [LIFE STAGE CEREMONY]. What do these rituals reveal about each tradition's understanding of [HUMAN NATURE/COMMUNITY/DIVINE RELATIONSHIP]?

8. **Theological Anthropology**: Compare how [RELIGION 1] and [RELIGION 2] understand [ASPECT OF HUMAN NATURE]. How do these different anthropological views affect their approaches to [ETHICS/SALVATION/COMMUNITY LIFE]?

9. **Scriptural Hermeneutics**: Examine how [RELIGIOUS TRADITION 1] and [RELIGIOUS TRADITION 2] interpret [CONTROVERSIAL PASSAGE/THEME]. What hermeneutical principles guide their different conclusions about [SPECIFIC ISSUE]?

10. **Divine Attributes Analysis**: Compare how [RELIGION 1] and [RELIGION 2] understand the divine attribute of [GOD'S CHARACTERISTIC]. What philosophical and experiential factors shape their different emphases?

11. **Religious Art and Symbolism**: Analyze the use of [RELIGIOUS SYMBOL/ARTISTIC MOTIF] in [TRADITION 1] versus [TRADITION 2]. How do cultural contexts and theological principles influence symbolic representation?

12. **Community Formation Patterns**: Compare how [RELIGIOUS MOVEMENT 1] and [RELIGIOUS MOVEMENT 2] developed their approach to [COMMUNITY STRUCTURE/MEMBERSHIP/IDENTITY]. What internal and external factors shaped their organizational evolution?

13. **Theodicy and Suffering**: Examine how [RELIGION 1] and [RELIGION 2] address the problem of [TYPE OF SUFFERING/EVIL]. What theological resources does each tradition offer for understanding and responding to [SPECIFIC CHALLENGE]?

14. **Eschatological Visions**: Compare how [TRADITION 1] and [TRADITION 2] envision [END TIMES/AFTERLIFE/ULTIMATE FULFILLMENT]. What do these different visions reveal about each tradition's

understanding of [HISTORY/HUMAN DESTINY/DIVINE PURPOSE]?

15. **Religious Innovation Response**: Analyze how [TRADITIONAL RELIGION] and [MODERN RELIGIOUS MOVEMENT] approach [CONTEMPORARY CHALLENGE]. What factors determine acceptance or rejection of [INNOVATION/ADAPTATION/REFORM]?

Textual Analysis and Interpretation Templates (15 prompts)

16. **Historical-Critical Analysis**: Apply historical-critical methodology to [SPECIFIC TEXT/PASSAGE], examining its [LITERARY FORM/HISTORICAL CONTEXT/THEOLOGICAL PURPOSE]. How does this analysis illuminate the text's meaning for both original and contemporary audiences?

17. **Literary Genre Exploration**: Analyze [SPECIFIC TEXT] as an example of [LITERARY GENRE]. How do the conventions of [GENRE TYPE] shape the text's message about [THEOLOGICAL THEME], and what interpretive challenges does this genre present?

18. **Redaction Criticism Application**: Examine [COMPOSITE TEXT] using redaction criticism to identify different editorial layers and theological perspectives. How do these multiple voices contribute to the text's final message about [CENTRAL THEME]?

19. **Canonical Context Analysis**: Analyze how [SPECIFIC TEXT] functions within the broader canonical context of [RELIGIOUS TRADITION]. How does its placement and intertextual relationships affect interpretation of [KEY PASSAGE/THEME]?

20. **Cultural Background Investigation**: Examine how understanding [ANCIENT CULTURAL PRACTICE/HISTORICAL CONTEXT] illuminates the meaning of [SPECIFIC PASSAGE]. What modern misunderstandings arise from ignorance of this background?

21. **Theological Theme Development**: Trace how the theme of [THEOLOGICAL CONCEPT] develops throughout [RELIGIOUS TEXT/CORPUS]. What theological evolution or consistency can be observed regarding [SPECIFIC ASPECT]?

22. **Feminist/Liberation Reading**: Apply [FEMINIST/LIBERATION/POSTCOLONIAL] hermeneutics to [SPECIFIC TEXT], focusing on [PARTICULAR CONCERN]. How does this interpretive lens reveal previously overlooked dimensions of the text's meaning?

23. **Comparative Literature Approach**: Compare [RELIGIOUS TEXT] with [CONTEMPORARY NON-RELIGIOUS TEXT] from the same cultural period, examining their treatment of [SHARED THEME]. What distinctive religious perspectives emerge from this comparison?

24. **Archaeological Correlation**: Examine how archaeological discoveries at [SPECIFIC SITE/CULTURE] correlate with or challenge the account in [RELIGIOUS TEXT]. What methodological principles should guide integration of textual and archaeological evidence?

25. **Translation Analysis**: Compare different translations of [SPECIFIC PASSAGE] in [RELIGIOUS TEXT], focusing on [CONTROVERSIAL WORD/PHRASE]. How do translation choices affect theological interpretation of [SPECIFIC DOCTRINE/PRACTICE]?

26. **Intertextual Analysis**: Examine how [LATER TEXT] uses or reinterprets [EARLIER TEXT], particularly regarding [SHARED THEME/MOTIF]. What theological development or continuity does this intertextual relationship reveal?

27. **Narrative Criticism Application**: Apply narrative criticism to [STORY/NARRATIVE TEXT], examining [PLOT/CHARACTER/SETTING] elements. How do these literary features contribute to the text's theological message about [CENTRAL THEME]?

28. **Wisdom Literature Analysis**: Analyze [WISDOM TEXT/PASSAGE] within the context of ancient Near Eastern wisdom traditions, focusing on its approach to [LIFE PROBLEM/ETHICAL ISSUE]. What distinctive theological insights emerge?

29. **Apocalyptic Literature Interpretation**: Examine [APOCALYPTIC TEXT] as a response to [HISTORICAL CRISIS/SITUATION], analyzing its use of [SYMBOLIC IMAGERY/TEMPORAL FRAMEWORK]. How should modern readers interpret its message about [DIVINE ACTION/HUMAN HOPE]?

30. **Devotional Literature Study**: Analyze [DEVOTIONAL/SPIRITUAL TEXT] as a guide for [SPIRITUAL PRACTICE/RELATIONSHIP WITH DIVINE], examining its understanding of [SPIRITUAL DEVELOPMENT/PRAYER/CONTEMPLATION]. How does it address [SPECIFIC SPIRITUAL CHALLENGE]?

Historical and Biographical Study Templates (15 prompts)

31. **Religious Figure Biography**: Analyze the life and impact of [RELIGIOUS LEADER/THINKER], focusing on their contribution to [SPECIFIC AREA]. How did [HISTORICAL CONTEXT/PERSONAL BACKGROUND] shape their approach to [RELIGIOUS ISSUE/INNOVATION]?

32. **Historical Period Analysis**: Examine the religious dimensions of [HISTORICAL PERIOD/EVENT], focusing on how [SPECIFIC RELIGIOUS TRADITION] responded to [CHALLENGE/OPPORTUNITY]. What long-term impacts did this period have on [RELIGIOUS DEVELOPMENT]?

33. **Institutional Development**: Trace the historical development of [RELIGIOUS INSTITUTION/PRACTICE] from its origins through [SPECIFIC TIME PERIOD]. What internal and external factors drove changes in [ORGANIZATIONAL STRUCTURE/THEOLOGICAL EMPHASIS]?

34. **Reform Movement Study**: Analyze the [REFORM/REVIVAL MOVEMENT] in [RELIGIOUS TRADITION], examining its response to [CONTEMPORARY CHALLENGE]. How did reformers balance [TRADITION] with [INNOVATION/ADAPTATION]?

35. **Persecution and Survival**: Examine how [RELIGIOUS COMMUNITY] survived and adapted during the period of [PERSECUTION/OPPRESSION]. What strategies enabled continuity of [FAITH/PRACTICE/COMMUNITY] under adverse conditions?

36. **Missionary Movement Analysis**: Analyze the [MISSIONARY MOVEMENT/EXPANSION] of [RELIGION] into [GEOGRAPHICAL/CULTURAL

REGION], focusing on [METHODS/ADAPTATION/RESISTANCE]. What factors determined success or failure in [CULTURAL TRANSLATION]?

37. **Schism and Division Study**: Examine the causes and consequences of the [RELIGIOUS SPLIT/SCHISM] between [GROUP 1] and [GROUP 2], focusing on disputes over [THEOLOGICAL/PRACTICAL ISSUE]. How did each side justify their position regarding [CENTRAL DISAGREEMENT]?

38. **Cultural Interaction Analysis**: Analyze the encounter between [RELIGION] and [CULTURE/PHILOSOPHY] during [TIME PERIOD], examining how [RELIGIOUS TRADITION] adapted to or resisted [CULTURAL INFLUENCE]. What theological innovations or tensions resulted?

39. **Women's Religious History**: Examine the role of women in [RELIGIOUS TRADITION] during [TIME PERIOD], focusing on [SPECIFIC ASPECT]. How did women navigate restrictions while exercising religious influence in [PARTICULAR AREA]?

40. **Social Justice Involvement**: Analyze how [RELIGIOUS LEADER/COMMUNITY] engaged with [SOCIAL JUSTICE ISSUE] during [HISTORICAL PERIOD]. What theological motivations drove their involvement, and what obstacles did they encounter?

41. **Religious Architecture Evolution**: Trace the development of [RELIGIOUS ARCHITECTURAL FORM] from [TIME PERIOD] to [TIME PERIOD], examining how changes reflect evolving theology and worship practices in [RELIGIOUS TRADITION].

42. **Sacred Text Development**: Examine the historical process by which [RELIGIOUS TEXT/CANON] reached its final form, focusing on [COMPILATION/EDITING/AUTHORIZATION]. What theological and political factors influenced these decisions?

43. **Conversion and Identity**: Analyze the conversion of [HISTORICAL FIGURE] from [ORIGINAL RELIGION] to [NEW RELIGION], examining the personal, social, and theological factors involved. How did this conversion impact [COMMUNITY/THOUGHT/PRACTICE]?

44. **Religious Conflict Resolution**: Examine how [RELIGIOUS COMMUNITIES] resolved the conflict over [DISPUTED ISSUE] during [TIME PERIOD]. What theological arguments and practical compromises enabled [RESOLUTION/COEXISTENCE]?

45. **Modernization Response**: Analyze how [TRADITIONAL RELIGIOUS INSTITUTION] responded to [MODERNIZING FORCE] during [TIME PERIOD]. What elements were preserved, adapted, or abandoned in the process of [MODERNIZATION/REFORM]?

Theological and Philosophical Inquiry Templates (15 prompts)

46. **Systematic Theology Construction**: Develop a systematic theological treatment of [DOCTRINE/CONCEPT] drawing from [RELIGIOUS TRADITION], addressing [CONTEMPORARY CHALLENGE]. How do traditional sources inform modern understanding of [SPECIFIC ISSUE]?

47. **Philosophical Dialogue Creation**: Construct a philosophical dialogue between [RELIGIOUS THINKER 1] and

[RELIGIOUS THINKER 2] on the topic of [PHILOSOPHICAL/THEOLOGICAL ISSUE]. How would each defend their position regarding [SPECIFIC DISAGREEMENT]?

48. **Theodicy Development**: Analyze the problem of [SPECIFIC TYPE OF EVIL/SUFFERING] from the perspective of [RELIGIOUS TRADITION], engaging with both classical theodicies and contemporary challenges. What resources does [TRADITION] offer for addressing [PARTICULAR DIFFICULTY]?

49. **Natural Theology Exploration**: Examine arguments for [RELIGIOUS CLAIM] based on [NATURAL EVIDENCE/REASON], evaluating both classical formulations and contemporary developments. How do modern [SCIENTIFIC/PHILOSOPHICAL] discoveries affect these arguments?

50. **Religious Epistemology**: Analyze how [RELIGIOUS TRADITION] understands religious knowledge and truth claims about [SPECIFIC DOCTRINE]. What roles do [REVELATION/REASON/EXPERIENCE/TRADITION] play in establishing [RELIGIOUS CERTAINTY]?

51. **Ethical Framework Application**: Apply the ethical framework of [RELIGIOUS TRADITION] to the contemporary issue of [MORAL DILEMMA]. How do core principles like [VIRTUE/DUTY/CONSEQUENCE] guide decision-making about [SPECIFIC ETHICAL CHALLENGE]?

52. **Divine Attribute Analysis**: Examine the logical coherence of affirming that God is both [DIVINE ATTRIBUTE 1] and [DIVINE ATTRIBUTE 2], addressing apparent contradictions in [SPECIFIC SCENARIO]. How do theologians resolve tensions between these attributes?

53. **Religious Language Philosophy**: Analyze how [RELIGIOUS TRADITION] understands religious language about [TRANSCENDENT REALITY], examining issues of [ANALOGY/SYMBOL/LITERAL MEANING]. What philosophical challenges arise in speaking about [DIVINE MYSTERY]?

54. **Interfaith Theological Dialogue**: Construct a theological dialogue between representatives of [RELIGION 1] and [RELIGION 2] on [SHARED CONCERN]. What common ground exists regarding

[ISSUE], and where do fundamental differences remain?

55. **Postmodern Challenge Response**: Analyze how [RELIGIOUS TRADITION] can respond to postmodern critiques of [TRUTH CLAIMS/AUTHORITY/GRAND NARRATIVES]. What aspects of [POSTMODERN THOUGHT] can be constructively engaged, and what must be resisted?

56. **Science and Religion Integration**: Examine how [RELIGIOUS TRADITION] can integrate insights from [SCIENTIFIC FIELD] regarding [NATURAL PHENOMENON]. What theological adjustments, if any, are required to accommodate [SCIENTIFIC DISCOVERY]?

57. **Religious Pluralism Position**: Develop a theological position on religious diversity from the perspective of [RELIGIOUS TRADITION], addressing the status of [OTHER RELIGIONS]. How does [TRADITION] balance [TRUTH CLAIMS] with [RESPECT FOR OTHERS]?

58. **Liberation Theology Application**: Apply liberation theological methodology to

[CONTEMPORARY OPPRESSION], using insights from [RELIGIOUS TRADITION]. How does [LIBERATION PERSPECTIVE] illuminate [SOCIAL INJUSTICE] and suggest [TRANSFORMATIVE ACTION]?

59. **Process Theology Integration**: Examine how process theological concepts might enrich understanding of [TRADITIONAL DOCTRINE] within [RELIGIOUS TRADITION]. What new insights emerge regarding [GOD'S NATURE/HUMAN RELATIONSHIP/COSMIC PURPOSE]?

60. **Mystical Experience Analysis**: Analyze accounts of [MYSTICAL EXPERIENCE] from [RELIGIOUS TRADITION], examining their epistemological status and theological significance. How should [EXTRAORDINARY EXPERIENCES] inform [DOCTRINE/PRACTICE]?

Contemporary Issues and Applied Ethics Templates (15 prompts)

61. **Bioethics Application**: Analyze [BIOETHICAL DILEMMA] from the perspective of [RELIGIOUS TRADITION], drawing on principles of [THEOLOGICAL CONCEPT]. How do competing values of

[VALUE 1] and [VALUE 2] require careful ethical discernment?

62. **Environmental Ethics Development**: Examine how [RELIGIOUS TRADITION] might address [ENVIRONMENTAL CRISIS/ISSUE], drawing on theological resources like [CREATION DOCTRINE/STEWARDSHIP/COSMIC SPIRITUALITY]. What practical steps follow from these theological convictions?

63. **Social Justice Analysis**: Apply the social justice principles of [RELIGIOUS TRADITION] to [CONTEMPORARY INEQUALITY], examining root causes and potential solutions. How do concepts like [HUMAN DIGNITY/PREFERENTIAL OPTION/PROPHETIC WITNESS] inform engagement with [SPECIFIC INJUSTICE]?

64. **Technology Ethics**: Analyze [TECHNOLOGICAL DEVELOPMENT] from the perspective of [RELIGIOUS TRADITION], examining implications for [HUMAN FLOURISHING/COMMUNITY/SPIRITUAL LIFE]. What ethical guidelines should govern use of [SPECIFIC TECHNOLOGY]?

65. **Economic Ethics Application**: Examine [ECONOMIC ISSUE/SYSTEM] through the lens of [RELIGIOUS TRADITION], evaluating it against principles of [JUSTICE/STEWARDSHIP/COMMUNITY]. What alternative approaches might better reflect [THEOLOGICAL VALUES]?

66. **Political Theology Development**: Analyze the proper relationship between [RELIGIOUS INSTITUTION] and [POLITICAL SYSTEM] regarding [POLICY ISSUE]. How should [RELIGIOUS PRINCIPLES] inform [POLITICAL ENGAGEMENT] without compromising [RELIGIOUS INTEGRITY]?

67. **Medical Ethics Integration**: Examine [MEDICAL PRACTICE/DECISION] from the perspective of [RELIGIOUS TRADITION], considering factors like [SANCTITY OF LIFE/QUALITY OF LIFE/PATIENT AUTONOMY]. How do religious and medical ethics interact in [CLINICAL SCENARIO]?

68. **Educational Philosophy**: Develop an educational philosophy based on [RELIGIOUS TRADITION] that addresses [CONTEMPORARY EDUCATIONAL CHALLENGE]. How do theological

concepts like [HUMAN FORMATION/TRUTH SEEKING/COMMUNITY BUILDING] inform [PEDAGOGICAL APPROACH]?

69. **Criminal Justice Reform**: Apply principles from [RELIGIOUS TRADITION] to [CRIMINAL JUSTICE ISSUE], examining alternatives to [PUNITIVE APPROACH]. How do concepts like [FORGIVENESS/RESTORATION/HUMAN DIGNITY] suggest [REFORM STRATEGIES]?

70. **Mental Health Integration**: Examine the relationship between [SPIRITUAL PRACTICE/BELIEF] and [MENTAL HEALTH ISSUE], analyzing both therapeutic benefits and potential problems. How can [RELIGIOUS RESOURCES] complement [PROFESSIONAL TREATMENT]?

71. **Immigration Ethics**: Analyze [IMMIGRATION POLICY/ISSUE] from the perspective of [RELIGIOUS TRADITION], drawing on principles like [HOSPITALITY/HUMAN DIGNITY/SOCIAL JUSTICE]. What

theological imperatives should guide [POLICY POSITIONS]?

72. **Media Ethics Application**: Examine [MEDIA ISSUE/TREND] through the lens of [RELIGIOUS TRADITION], considering impacts on [HUMAN RELATIONSHIPS/TRUTH TELLING/COMMUNITY VALUES]. How should religious communities engage with [MEDIA TECHNOLOGY/CONTENT]?

73. **Business Ethics Integration**: Apply principles from [RELIGIOUS TRADITION] to [BUSINESS PRACTICE/DECISION], examining tensions between [PROFIT] and [ETHICAL VALUES]. How can [RELIGIOUS PRINCIPLES] inform [CORPORATE RESPONSIBILITY]?

74. **Family Ethics Analysis**: Examine [FAMILY ISSUE/STRUCTURE] from the perspective of [RELIGIOUS TRADITION], considering how [SOCIAL CHANGES] affect [TRADITIONAL VALUES]. How should [RELIGIOUS COMMUNITIES] respond to [CHANGING FAMILY FORMS]?

75. **Global Ethics Application**: Analyze [GLOBAL ISSUE/CHALLENGE] through the lens of [RELIGIOUS TRADITION], examining responsibilities toward [GLOBAL COMMUNITY]. How do concepts like [UNIVERSAL SOLIDARITY/GLOBAL STEWARDSHIP] inform responses to [INTERNATIONAL PROBLEMS]?

How to Use These Customizable Prompts

Substitution Guidelines:

[RELIGION/TRADITION]: Insert specific religious traditions (Christianity, Islam, Judaism, Buddhism, Hinduism, Indigenous traditions, etc.)

[RELIGIOUS FIGURE]: Insert names of religious leaders, theologians, mystics, reformers, or other significant figures

[SACRED TEXT]: Insert specific scriptures, commentaries, or religious writings

[THEOLOGICAL CONCEPT]: Insert doctrines, beliefs, or religious ideas (salvation, divine nature, afterlife, etc.)

[HISTORICAL PERIOD]: Insert specific time periods, eras, or historical events

[CONTEMPORARY ISSUE]: Insert current challenges, debates, or social questions

[ETHICAL DILEMMA]: Insert specific moral challenges or bioethical issues

[CULTURAL CONTEXT]: Insert geographical regions, ethnic groups, or cultural settings

Academic Enhancement Tips:

1. **Research Depth**: Use each customized prompt as a foundation for extensive research using primary sources, scholarly articles, and academic databases

2. **Comparative Analysis**: Many prompts work well when comparing 2-3 different religions, time periods, or perspectives

3. **Interdisciplinary Approach**: Connect religious studies with other fields like history, philosophy, sociology, psychology, or ethics

4. **Critical Thinking**: Always engage multiple viewpoints and avoid oversimplification of complex religious phenomena

5. **Contemporary Relevance**: Connect historical or theological analysis with current applications and challenges

Suggested Applications:

- **Research Papers**: Develop 15-25 page research projects
- **Class Presentations**: Create focused 10-15 minute presentations
- **Discussion Posts**: Generate thoughtful online discussion contributions
- **Exam Preparation**: Practice comprehensive essay responses
- **Thesis Development**: Use as starting points for senior thesis topics
- **Independent Study**: Structure self-directed learning projects

These templates provide maximum flexibility while maintaining academic rigor appropriate for undergraduate religious studies research and analysis.

75 AI Prompts for Bachelor of Divinity and Theology Students

Research, Inquiry, and Analysis at the Undergraduate Level

75 AI Prompts for Bachelor of Divinity and Theology Students

Research, Inquiry, and Analysis at the Undergraduate Level

A comprehensive collection of 75 AI prompts specifically designed for Bachelor of Divinity and Theology students. These prompts are organized into five key areas:

1. **Biblical Studies & Exegesis** (15 prompts) - covering hermeneutics, textual analysis, and comparative interpretation

2. **Church History** (15 prompts) - spanning from early church through contemporary movements

3. **Systematic Theology** (15 prompts) - exploring major doctrines and theological methods

4. **Ethics and Moral Theology** (15 prompts) - addressing contemporary ethical challenges from theological perspectives

5. **Practical Theology and Ministry** (15 prompts) - connecting theological knowledge to ministerial practice

Each prompt is designed to:

- Encourage critical thinking and analysis
- Promote engagement with primary sources
- Foster comparative theological study
- Connect historical theology to contemporary issues
- Develop skills in theological argumentation
- Prepare students for advanced theological study

The prompts can be adapted for research papers, class discussions, examinations, or personal study. They're structured to help students move beyond surface-level understanding to engage deeply with theological concepts, methods, and applications at the undergraduate level.

75 AI Prompts for Bachelor of Divinity and Theology Students

Research, Inquiry, and Analysis at the Undergraduate Level

Biblical Studies & Exegesis (15 Prompts)

1. **Textual Analysis**: Analyze the literary structure and theological themes in [specific biblical passage]. What hermeneutical approaches best illuminate its meaning for contemporary readers?

2. **Comparative Exegesis**: Compare how three different theological traditions (Catholic, Protestant, Orthodox) interpret [specific biblical text]. What are the key interpretive differences and their theological implications?

3. **Historical-Critical Method**: Apply historical-critical methodology to examine the authorship, dating, and historical context of [biblical book]. How does this analysis affect theological interpretation?

4. **Genre Analysis**: Examine the literary genre of [biblical text] and explain how

understanding its genre influences theological interpretation and application.

5. **Intertextuality Study**: Trace the intertextual connections between [Old Testament passage] and its use in [New Testament passage]. What theological significance emerges from this relationship?

6. **Translation Comparison**: Compare three different English translations of [specific verse] with the original Hebrew/Greek. What theological implications arise from translation choices?

7. **Canonical Criticism**: Analyze how [specific biblical book] functions within the broader canonical context. What unique theological contribution does it make to Scripture as a whole?

8. **Narrative Theology**: Examine the theological significance of narrative elements in [biblical story]. How does the story structure convey theological meaning?
9. **Prophetic Literature Analysis**: Investigate the socio-political context of [prophetic book] and analyze how the prophet's message addresses contemporary issues of justice and covenant faithfulness.

10. **Wisdom Literature Study**: Compare the theological perspectives on suffering found in Job, Ecclesiastes, and Proverbs. What unified and divergent themes emerge?

11. **Pauline Theology**: Analyze Paul's understanding of [theological concept] across his epistles. How does his thought develop or remain consistent?

12. **Gospel Comparison**: Examine how the four Gospels present [specific aspect of Jesus' ministry]. What unique theological emphasis does each evangelist contribute?

13. **Apocalyptic Literature**: Decode the symbolic language in [apocalyptic text] and explain its theological message for its original audience and contemporary readers.

14. **Covenant Theology**: Trace the development of covenant concepts from Genesis through Revelation. How do different covenant relationships reveal God's character?

15. **Biblical Ethics**: Analyze the ethical teachings in [biblical passage] and discuss their application to contemporary moral dilemmas.

Church History (15 Prompts)

16. **Patristic Development**: Examine how [Church Father] contributed to the development of [theological doctrine]. What were the historical circumstances that shaped their thinking?

17. **Ecumenical Councils**: Analyze the theological and political factors that led to [specific council] and evaluate its lasting impact on Christian doctrine.

18. **Reformation Analysis**: Compare Martin Luther's and John Calvin's positions on [theological issue]. How did their historical contexts influence their theological differences?
19. **Early Church Persecution**: Investigate how persecution under [Roman emperor] shaped early Christian theology and practice. What theological themes emerged from this experience?

20. **Monasticism Study**: Examine the theological foundations of [monastic movement] and its influence on medieval Christian spirituality and scholarship.

21. **Medieval Scholasticism**: Compare Thomas Aquinas and Duns Scotus on [theological

question]. How did their philosophical backgrounds influence their theological conclusions?

22. **Missionary Movement**: Analyze the theological motivations and methods of [missionary figure/movement]. What were the theological and cultural implications of their work?

23. **Denominational Origins**: Trace the theological and historical factors that led to the formation of [denomination]. What distinctive theological emphases emerged?

24. **Women in Church History**: Examine the contributions of [female theologian/mystic] to Christian thought. How did gender influence their theological expression and reception?

25. **Eastern Christianity**: Compare Eastern Orthodox and Western Catholic approaches to [theological issue] during [historical period]. What cultural factors influenced these differences?

26. **Revival Movements**: Analyze the theological characteristics of [revival movement]. What social and spiritual factors contributed to its emergence and impact?

27. **Modern Theology Origins**: Examine how [historical event/period] influenced the development of modern theological thought. What new questions and methods emerged?

28. **Liberation Theology**: Trace the historical development of liberation theology in [region/context]. What theological and social factors contributed to its emergence?

29. **Ecumenical Movement**: Analyze the theological foundations of the modern ecumenical movement. What theological convergences and divergences have been identified?

30. **Contextual Christianity**: Examine how Christianity adapted to [cultural context] during [historical period]. What theological indigenization occurred?

Systematic Theology (15 Prompts)

31. **Doctrine of God**: Compare classical theism, process theology, and open theism on the nature of divine immutability. What are the biblical and philosophical arguments for each position?

32. **Christology**: Analyze the Chalcedonian definition of Christ's two natures. How does

this formulation address contemporary Christological questions?

33. **Pneumatology**: Examine different theological traditions' understanding of the Holy Spirit's role in sanctification. What are the pastoral and liturgical implications?

34. **Anthropology**: Compare Augustinian and Pelagian views of human nature and original sin. How do these perspectives influence soteriology and ethics?

35. **Soteriology**: Analyze the theological differences between justification and sanctification across various traditions. What is the relationship between faith and works?

36. **Ecclesiology**: Examine different models of church authority (episcopal, presbyterian, congregational). What biblical and theological arguments support each model?

37. **Eschatology**: Compare premillennial, postmillennial, and amillennial interpretations of biblical prophecy. What hermeneutical principles underlie each approach?

38. **Trinity**: Analyze the theological development from economic Trinity to immanent Trinity. How do contemporary theologians address traditional formulations?
39. **Scripture and Tradition**: Examine the relationship between biblical authority and church tradition in different theological systems. What are the epistemological implications?

40. **Natural Theology**: Analyze the role of natural revelation in theological knowledge. How do different traditions balance general and special revelation?

41. **Theodicy**: Examine different theological responses to the problem of evil. How do they maintain both divine goodness and sovereignty?

42. **Sacramental Theology**: Compare Protestant, Catholic, and Orthodox understandings of [specific sacrament]. What theological principles underlie these differences?

43. **Divine Providence**: Analyze the tension between divine sovereignty and human freedom. How do different theological systems resolve this paradox?

44. **Theological Method**: Examine the relationship between faith and reason in theological inquiry. What role should philosophy play in theology?

45. **Contemporary Issues**: Apply traditional theological categories to [modern ethical issue]. What theological resources are most relevant?

Ethics and Moral Theology (15 Prompts)

46. **Bioethics**: Analyze competing theological perspectives on [bioethical issue]. What biblical and theological principles should guide Christian decision-making?

47. **Social Justice**: Examine the theological foundations for Christian engagement with [social issue]. How do different traditions approach social responsibility?

48. **Sexual Ethics**: Compare traditional and progressive Christian approaches to [sexual ethics topic]. What hermeneutical and theological factors influence these positions?

49. **Economic Ethics**: Analyze Christian perspectives on capitalism, socialism, and mixed economies. What theological principles should guide economic systems?

50. **Environmental Ethics**: Examine theological approaches to creation care and environmental stewardship. How does Christian anthropology inform ecological responsibility?

51. **War and Peace**: Compare pacifist, just war, and crusade/holy war theories from theological perspectives. What biblical and

theological arguments support each position?

52. **Medical Ethics**: Analyze the theological dimensions of [medical ethics issue]. How should Christian healthcare professionals navigate ethical dilemmas?

53. **Business Ethics**: Examine Christian approaches to [business ethics topic]. What theological principles should guide marketplace behavior?

54. **Political Theology**: Analyze the relationship between church and state from different theological perspectives. What is the proper Christian approach to political engagement?

55. **Human Rights**: Examine theological foundations for human dignity and rights. How do different Christian traditions ground universal human rights?

56. **Technology Ethics**: Analyze the theological implications of [emerging technology]. What ethical frameworks should guide Christian responses?

57. **Immigration Ethics**: Examine biblical and theological perspectives on immigration and

refugee care. What obligations do Christians have toward migrants?

58. **Criminal Justice**: Analyze Christian approaches to punishment, rehabilitation, and restorative justice. What theological principles should guide criminal justice reform?

59. **Virtue Ethics**: Compare Aristotelian virtue ethics with Christian virtue traditions. How do theological virtues relate to classical virtues?

60. **Comparative Religious Ethics**: Examine Christian responses to ethical dilemmas in pluralistic societies. How should Christians engage with other moral traditions?

Practical Theology and Ministry (15 Prompts)

61. **Homiletics**: Analyze the theological and rhetorical elements of effective preaching. How should biblical exegesis inform sermon preparation and delivery?

62. **Pastoral Care**: Examine theological approaches to [pastoral care situation]. What theological resources guide effective pastoral ministry?

63. **Liturgical Theology**: Analyze the theological significance of [liturgical element/season]. How does worship shape theological understanding and spiritual formation?

64. **Christian Education**: Examine theological foundations for Christian formation across different life stages. What pedagogical approaches best serve theological education?

65. **Youth Ministry**: Analyze the theological and developmental considerations for adolescent faith formation. What unique challenges and opportunities exist?

66. **Missiology**: Examine theological approaches to cross-cultural evangelism and

church planting. How should gospel contextualization preserve theological integrity?

67. **Interfaith Dialogue**: Analyze Christian theological approaches to dialogue with [specific religion]. What theological principles should guide interfaith engagement?

68. **Congregational Leadership**: Examine biblical and theological models of church leadership. How should theology inform pastoral authority and decision-making?

69. **Worship and Music**: Analyze the theological dimensions of congregational music and worship styles. What criteria should guide worship planning?
70. **Church Planting**: Examine theological and practical considerations for establishing new congregations. What ecclesiological principles should guide church development?

71. **Spiritual Direction**: Analyze theological approaches to spiritual mentoring and discipleship. How do different traditions understand spiritual guidance?

72. **Crisis Ministry**: Examine theological responses to [type of crisis/trauma]. What theological resources help communities navigate suffering?

73. **Digital Ministry**: Analyze the theological implications of online worship, virtual community, and digital discipleship. How do technological mediations affect ministry?

74. **Urban Ministry**: Examine theological approaches to ministry in urban contexts. What unique challenges and opportunities exist for city churches?

75. **Rural Ministry**: Analyze the theological and practical considerations for rural and small-town ministry. How should geographic context influence ministerial approach?

How to Use These Prompts Effectively

For Research Papers:

- Use prompts as starting points for deeper investigation
- Develop thesis statements from prompt responses
- Create bibliographies from suggested sources and perspectives

For Class Discussions:

- Adapt prompts for group dialogue and debate
- Use comparative prompts to explore different viewpoints
- Encourage critical analysis rather than simple answers

For Personal Study:

- Use prompts for journal reflection and spiritual formation
- Apply theological concepts to contemporary situations
- Develop personal theological positions through careful analysis

For Examination Preparation:

- Practice articulating theological positions clearly
- Compare and contrast different theological perspectives

- Apply theological knowledge to practical scenarios

Research Tips:

- Always consult primary sources alongside secondary literature
- Engage with diverse theological traditions and perspectives
- Consider historical, cultural, and linguistic contexts
- Apply appropriate hermeneutical and methodological approaches
- Connect theological study to practical ministry and Christian living

75 Customizable AI Prompts for Bachelor of Divinity and Theology Students

75 Customizable AI Prompts for Bachelor of Divinity and Theology Students

Bachelor of Divinity and Theology Prompts

These 75 customizable prompts for Bachelor of Divinity and Theology students are designed to facilitate deep theological inquiry across five core areas: Biblical Studies & Exegesis, Church History, Systematic Theology, Ethics and Moral Theology, and Practical Theology and Ministry. Each prompt contains strategic placeholder brackets that allow students and instructors to insert specific topics, making them adaptable for diverse research interests, denominational perspectives, and contemporary issues.

The significance of these prompts lies in their ability to transform theological education by providing structured frameworks for critical analysis while maintaining flexibility for personalized learning. They help students develop essential skills including rigorous exegetical methodology, historical-critical thinking, systematic theological reasoning, ethical decision-making, and practical ministry application. By using these prompts, students learn to engage primary sources, compare theological traditions, apply ancient wisdom to contemporary challenges, and articulate coherent theological positions—ultimately preparing them for advanced seminary work, pastoral ministry, or academic careers in religious studies.

75 Customizable AI Prompts for Bachelor of Divinity and Theology Students

Research, Inquiry, and Analysis at the Undergraduate Level

Biblical Studies & Exegesis (15 Prompts)

1. **Textual Analysis**: Analyze the literary structure and theological themes in [BIBLICAL PASSAGE/BOOK]. What hermeneutical approaches best illuminate its meaning for [TARGET AUDIENCE/CONTEMPORARY ISSUE]?

2. **Comparative Exegesis**: Compare how [NUMBER] different theological traditions ([TRADITION 1], [TRADITION 2], [TRADITION 3]) interpret [BIBLICAL TEXT]. What are the key interpretive

differences and their implications for [THEOLOGICAL TOPIC].

3. **Historical-Critical Method**: Apply historical-critical methodology to examine the authorship, dating, and historical context of [BIBLICAL BOOK]. How does this analysis affect theological interpretation of [SPECIFIC THEME/DOCTRINE]?
4. **Genre Analysis**: Examine the literary genre of [BIBLICAL TEXT] and explain how understanding its genre influences theological interpretation and application to [CONTEMPORARY SITUATION].

5. **Intertextuality Study**: Trace the intertextual connections between [OLD TESTAMENT PASSAGE] and its use in [NEW TESTAMENT PASSAGE]. What theological significance emerges regarding [THEOLOGICAL CONCEPT]?

6. **Translation Comparison**: Compare [NUMBER] different English translations of [SPECIFIC VERSE/PASSAGE] with the original [HEBREW/GREEK]. What theological implications arise from translation choices affecting [DOCTRINAL UNDERSTANDING]?

7. **Canonical Criticism**: Analyze how [BIBLICAL BOOK] functions within the broader canonical context. What unique theological contribution does it make to understanding [THEOLOGICAL THEME]?

8. **Narrative Theology**: Examine the theological significance of narrative elements in [BIBLICAL STORY]. How does the story structure convey theological meaning about [GOD'S CHARACTER/HUMAN NATURE/SALVATION]?

9. **Prophetic Literature Analysis**: Investigate the socio-political context of [PROPHETIC BOOK] and analyze how the prophet's message addresses [CONTEMPORARY ISSUE] through themes of [JUSTICE/COVENANT/JUDGMENT].

10. **Wisdom Literature Study**: Compare the theological perspectives on [THEME/TOPIC] found in [WISDOM BOOK 1], [WISDOM BOOK 2], and [WISDOM BOOK 3]. What unified and divergent themes emerge regarding [THEOLOGICAL QUESTION]?

11. **Pauline Theology**: Analyze Paul's understanding of [THEOLOGICAL

CONCEPT] across [SPECIFIC EPISTLES]. How does his thought develop or remain consistent regarding [DOCTRINAL ISSUE]?

12. **Gospel Comparison**: Examine how the four Gospels present [ASPECT OF JESUS' MINISTRY/TEACHING]. What unique theological emphasis does each evangelist contribute to understanding [CHRISTOLOGICAL THEME]?

13. **Apocalyptic Literature**: Decode the symbolic language in [APOCALYPTIC TEXT] and explain its theological message for [ORIGINAL AUDIENCE] and contemporary [TARGET GROUP] facing [CURRENT CHALLENGE].

14. **Covenant Theology**: Trace the development of [COVENANT CONCEPT] from [STARTING POINT] through [ENDING POINT]. How do different covenant relationships reveal [ASPECT OF GOD'S CHARACTER]?

15. **Biblical Ethics**: Analyze the ethical teachings in [BIBLICAL PASSAGE] and discuss their application to contemporary [MORAL DILEMMA/ETHICAL ISSUE].

Church History (15 Prompts)

16. **Patristic Development**: Examine how [CHURCH FATHER] contributed to the development of [THEOLOGICAL DOCTRINE] during [TIME PERIOD]. What were the [HISTORICAL CIRCUMSTANCES] that shaped their thinking?

17. **Ecumenical Councils**: Analyze the theological and political factors that led to [SPECIFIC COUNCIL] and evaluate its lasting impact on [CHRISTIAN DOCTRINE/CHURCH PRACTICE].

18. **Reformation Analysis**: Compare [REFORMER 1]'s and [REFORMER 2]'s positions on [THEOLOGICAL ISSUE]. How did their [HISTORICAL CONTEXTS] influence their theological differences regarding [SPECIFIC DOCTRINE]?

19. **Early Church Persecution**: Investigate how persecution under [ROMAN EMPEROR/HISTORICAL PERIOD] shaped early Christian [THEOLOGY/PRACTICE]. What theological themes emerged from this experience of [SUFFERING/MARTYRDOM]?

20. **Monasticism Study**: Examine the theological foundations of [MONASTIC MOVEMENT/ORDER] and its influence on [MEDIEVAL PERIOD] Christian spirituality and [SCHOLARSHIP/PRACTICE].

21. **Medieval Scholasticism**: Compare [SCHOLASTIC THEOLOGIAN 1] and [SCHOLASTIC THEOLOGIAN 2] on [THEOLOGICAL QUESTION]. How did their [PHILOSOPHICAL BACKGROUNDS] influence their theological conclusions about [DOCTRINE]?

22. **Missionary Movement**: Analyze the theological motivations and methods of [MISSIONARY FIGURE/MOVEMENT] in [GEOGRAPHIC REGION/TIME PERIOD]. What were the theological and cultural implications for [TARGET CULTURE/CHURCH DEVELOPMENT]?

23. **Denominational Origins**: Trace the theological and historical factors that led to the formation of [DENOMINATION]. What distinctive theological emphases emerged regarding [CHURCH PRACTICE/DOCTRINE]?

24. **Women in Church History**: Examine the contributions of [FEMALE THEOLOGIAN/MYSTIC/LEADER] to Christian thought during [TIME PERIOD]. How did [GENDER/SOCIAL CONTEXT] influence their theological expression and reception?

25. **Eastern Christianity**: Compare Eastern Orthodox and Western Catholic approaches to [THEOLOGICAL ISSUE] during [HISTORICAL PERIOD]. What [CULTURAL/PHILOSOPHICAL] factors influenced these differences?

26. **Revival Movements**: Analyze the theological characteristics of [REVIVAL MOVEMENT] in [GEOGRAPHIC LOCATION/TIME PERIOD]. What [SOCIAL/SPIRITUAL] factors contributed to its emergence and impact on [CHURCH/SOCIETY]?

27. **Modern Theology Origins**: Examine how [HISTORICAL EVENT/PERIOD] influenced the development of [THEOLOGICAL MOVEMENT]. What new questions and methods emerged regarding [THEOLOGICAL TOPIC]?

28. **Liberation Theology**: Trace the historical development of [LIBERATION THEOLOGY VARIANT] in [REGION/CONTEXT]. What [THEOLOGICAL/SOCIAL] factors contributed to its emergence and approach to [SOCIAL ISSUE]?

29. **Ecumenical Movement**: Analyze the theological foundations of [ECUMENICAL INITIATIVE/ORGANIZATION]. What theological convergences and divergences have been identified regarding [DENOMINATIONAL DIFFERENCES]?

30. **Contextual Christianity**: Examine how Christianity adapted to [CULTURAL CONTEXT] during [HISTORICAL PERIOD]. What theological indigenization occurred in [WORSHIP/THEOLOGY/PRACTICE]?

Systematic Theology (15 Prompts)

31. **Doctrine of God**: Compare [THEOLOGICAL POSITION 1], [THEOLOGICAL POSITION 2], and [THEOLOGICAL POSITION 3] on [DIVINE ATTRIBUTE]. What are the biblical and philosophical arguments for

each position regarding [THEOLOGICAL IMPLICATION]?

32. **Christology**: Analyze [CHRISTOLOGICAL FORMULATION/COUNCIL] definition of [CHRISTOLOGICAL ISSUE]. How does this formulation address [CONTEMPORARY CHRISTOLOGICAL QUESTIONS]?

33. **Pneumatology**: Examine different theological traditions' understanding of the Holy Spirit's role in [THEOLOGICAL PROCESS/CHRISTIAN EXPERIENCE]. What are the [PASTORAL/LITURGICAL] implications for [CHURCH PRACTICE]?

34. **Anthropology**: Compare [THEOLOGICAL POSITION 1] and [THEOLOGICAL POSITION 2] views of [ASPECT OF HUMAN NATURE]. How do these perspectives influence [SOTERIOLOGY/ETHICS/PASTORAL CARE]?

35. **Soteriology**: Analyze the theological differences between [SOTERIOLOGICAL CONCEPT 1] and [SOTERIOLOGICAL CONCEPT 2] across [THEOLOGICAL TRADITIONS]. What is the relationship

between [FAITH/WORKS/GRACE]?

36. **Ecclesiology**: Examine different models of [CHURCH STRUCTURE/AUTHORITY] ([MODEL 1], [MODEL 2], [MODEL 3]). What biblical and theological arguments support each model for [CHURCH GOVERNANCE/MINISTRY]?

37. **Eschatology**: Compare [ESCHATOLOGICAL POSITION 1], [ESCHATOLOGICAL POSITION 2], and [ESCHATOLOGICAL POSITION 3] interpretations of [PROPHETIC/ESCHATOLOGICAL TOPIC]. What hermeneutical principles underlie each approach to [END TIMES EVENTS]?

38. **Trinity**: Analyze the theological development from [TRINITARIAN CONCEPT 1] to [TRINITARIAN CONCEPT 2]. How do contemporary theologians address [TRADITIONAL FORMULATION/CHALLENGE] regarding [TRINITARIAN DOCTRINE]?

39. **Scripture and Tradition**: Examine the relationship between [BIBLICAL AUTHORITY] and [CHURCH TRADITION] in [THEOLOGICAL

SYSTEM/DENOMINATION]. What are the epistemological implications for [THEOLOGICAL METHOD]?

40. **Natural Theology**: Analyze the role of [NATURAL REVELATION] in theological knowledge within [THEOLOGICAL TRADITION]. How does this tradition balance [GENERAL REVELATION] and [SPECIAL REVELATION] regarding [KNOWLEDGE OF GOD]?

41. **Theodicy**: Examine [THEOLOGICAL RESPONSE] to the problem of [EVIL/SUFFERING]. How does this approach maintain both [DIVINE ATTRIBUTE 1] and [DIVINE ATTRIBUTE 2]?

42. **Sacramental Theology**: Compare [TRADITION 1], [TRADITION 2], and [TRADITION 3] understandings of [SPECIFIC SACRAMENT]. What theological principles underlie these differences in [SACRAMENTAL PRACTICE]?

43. **Divine Providence**: Analyze the tension between [DIVINE SOVEREIGNTY] and [HUMAN FREEDOM] in [THEOLOGICAL SYSTEM]. How does

this system resolve the paradox regarding [THEOLOGICAL IMPLICATION]?

44. **Theological Method**: Examine the relationship between [FAITH] and [REASON] in [THEOLOGICAL APPROACH/THEOLOGIAN]. What role should [PHILOSOPHY/EXPERIENCE/TRADITION] play in [THEOLOGICAL INQUIRY]?

45. **Contemporary Issues**: Apply traditional theological categories to [MODERN ETHICAL/SOCIAL ISSUE]. What theological resources from [THEOLOGICAL TRADITION/DOCTRINE] are most relevant to [CONTEMPORARY CHALLENGE]?

Ethics and Moral Theology (15 Prompts)

46. **Bioethics**: Analyze competing theological perspectives on [BIOETHICAL ISSUE] from [TRADITION 1] and [TRADITION 2] viewpoints. What [BIBLICAL/THEOLOGICAL] principles should guide Christian decision-making regarding [MEDICAL/LIFE ISSUE]?

47. **Social Justice**: Examine the theological foundations for Christian engagement with [SOCIAL ISSUE] in [CULTURAL CONTEXT]. How do different traditions approach social responsibility regarding [JUSTICE CONCERN]?

48. **Sexual Ethics**: Compare [TRADITIONAL/PROGRESSIVE] Christian approaches to [SEXUAL ETHICS TOPIC]. What [HERMENEUTICAL/THEOLOGICAL] factors influence these positions on [HUMAN SEXUALITY ISSUE]?

49. **Economic Ethics**: Analyze Christian perspectives on [ECONOMIC SYSTEM/ISSUE] from [THEOLOGICAL TRADITION]. What theological principles should guide [ECONOMIC DECISIONS/POLICY] regarding [WEALTH/POVERTY/JUSTICE]?

50. **Environmental Ethics**: Examine theological approaches to [ENVIRONMENTAL ISSUE] and [CREATION CARE]. How does Christian [ANTHROPOLOGY/CREATION THEOLOGY] inform ecological responsibility regarding [ENVIRONMENTAL CHALLENGE]?

51. **War and Peace**: Compare [PACIFIST], [JUST WAR], and [OTHER] theories from [THEOLOGICAL PERSPECTIVE]. What biblical and theological arguments support each position regarding [CONFLICT/VIOLENCE SITUATION]?

52. **Medical Ethics**: Analyze the theological dimensions of [MEDICAL ETHICS ISSUE] in [HEALTHCARE CONTEXT]. How should Christian healthcare professionals navigate ethical dilemmas involving [MEDICAL DECISION/PATIENT CARE]?

53. **Business Ethics**: Examine Christian approaches to [BUSINESS ETHICS TOPIC] in [MARKETPLACE CONTEXT]. What theological principles should guide [BUSINESS PRACTICE/DECISION] regarding [ETHICAL DILEMMA]?

54. **Political Theology**: Analyze the relationship between [CHURCH] and [STATE] from [THEOLOGICAL PERSPECTIVE]. What is the proper Christian approach to [POLITICAL ENGAGEMENT/CIVIC RESPONSIBILITY] in [POLITICAL CONTEXT]?

55. **Human Rights**: Examine theological foundations for [HUMAN DIGNITY ASPECT] and [RIGHTS ISSUE]. How do different Christian traditions ground [UNIVERSAL HUMAN RIGHTS] in [THEOLOGICAL DOCTRINE]?

56. **Technology Ethics**: Analyze the theological implications of [EMERGING TECHNOLOGY] for [HUMAN FLOURISHING/COMMUNITY]. What ethical frameworks should guide Christian responses to [TECHNOLOGICAL CHALLENGE]?

57. **Immigration Ethics**: Examine biblical and theological perspectives on [IMMIGRATION/REFUGEE ISSUE]. What obligations do Christians have toward [MIGRANT POPULATION] based on [BIBLICAL/THEOLOGICAL PRINCIPLE]?

58. **Criminal Justice**: Analyze Christian approaches to [PUNISHMENT/REHABILITATION/RESTORATIVE JUSTICE]. What theological principles should guide [CRIMINAL JUSTICE REFORM] regarding [JUSTICE ISSUE]?

59. **Virtue Ethics**: Compare [CLASSICAL VIRTUE ETHICS] with [CHRISTIAN VIRTUE TRADITION] regarding [MORAL VIRTUE/CHARACTER]. How do [THEOLOGICAL VIRTUES] relate to [CLASSICAL VIRTUES] in [MORAL FORMATION]?

60. **Comparative Religious Ethics**: Examine Christian responses to [ETHICAL DILEMMA] in [PLURALISTIC SOCIETY]. How should Christians engage with [OTHER MORAL TRADITION] regarding [SHARED ETHICAL CONCERN]?

Practical Theology and Ministry (15 Prompts)

61. **Homiletics**: Analyze the theological and rhetorical elements of effective preaching on [BIBLICAL TEXT/TOPIC]. How should [EXEGETICAL METHOD] inform sermon preparation for [TARGET CONGREGATION/OCCASION]?

62. **Pastoral Care**: Examine theological approaches to [PASTORAL CARE SITUATION] in [MINISTRY CONTEXT]. What theological resources guide effective pastoral ministry to [SPECIFIC NEED/CRISIS]?

63. **Liturgical Theology**: Analyze the theological significance of [LITURGICAL ELEMENT/SEASON] in [WORSHIP TRADITION]. How does [WORSHIP PRACTICE] shape theological understanding of [THEOLOGICAL CONCEPT]?

64. **Christian Education**: Examine theological foundations for Christian formation of [TARGET AGE GROUP/DEMOGRAPHIC]. What [PEDAGOGICAL APPROACHES] best

serve theological education regarding [FAITH FORMATION GOAL]?

65. **Youth Ministry**: Analyze the theological and developmental considerations for [ADOLESCENT FAITH ISSUE]. What unique challenges and opportunities exist for [YOUTH MINISTRY APPROACH] in [CULTURAL CONTEXT]?

66. **Missiology**: Examine theological approaches to [CROSS-CULTURAL MINISTRY] in [GEOGRAPHIC/CULTURAL CONTEXT]. How should [GOSPEL CONTEXTUALIZATION] preserve theological integrity while addressing [CULTURAL CONSIDERATIONS]?
67. **Interfaith Dialogue**: Analyze Christian theological approaches to dialogue with [SPECIFIC RELIGION/WORLDVIEW]. What theological principles should guide [INTERFAITH ENGAGEMENT] regarding [SHARED CONCERN/DIFFERENCE]?

68. **Congregational Leadership**: Examine biblical and theological models of [CHURCH LEADERSHIP ASPECT] in [CONGREGATIONAL CONTEXT]. How should theology inform [PASTORAL

AUTHORITY/DECISION-MAKING] regarding [LEADERSHIP CHALLENGE]?

69. **Worship and Music**: Analyze the theological dimensions of [WORSHIP ELEMENT/MUSICAL STYLE] in [CONGREGATIONAL SETTING]. What criteria should guide [WORSHIP PLANNING] for [TARGET CONGREGATION]?

70. **Church Planting**: Examine theological and practical considerations for [CHURCH PLANTING CONTEXT] in [GEOGRAPHIC/DEMOGRAPHIC AREA]. What [ECCLESIOLOGICAL PRINCIPLES] should guide [CHURCH DEVELOPMENT] for [TARGET COMMUNITY]?

71. **Spiritual Direction**: Analyze theological approaches to [SPIRITUAL MENTORING/DISCIPLESHIP] in [MINISTRY CONTEXT]. How do different traditions understand [SPIRITUAL GUIDANCE] for [SPIRITUAL GROWTH AREA]?

72. **Crisis Ministry**: Examine theological responses to [TYPE OF CRISIS/TRAUMA] in [COMMUNITY CONTEXT]. What

theological resources help [AFFECTED COMMUNITY] navigate [SUFFERING/CHALLENGE]?

73. **Digital Ministry**: Analyze the theological implications of [DIGITAL MINISTRY APPROACH] for [CHURCH COMMUNITY]. How do [TECHNOLOGICAL MEDIATIONS] affect [MINISTRY PRACTICE/COMMUNITY FORMATION]?

74. **Urban Ministry**: Examine theological approaches to ministry in [URBAN CONTEXT/CHALLENGE]. What unique opportunities exist for [CITY CHURCHES] to address [URBAN ISSUE]?

75. **Rural Ministry**: Analyze the theological and practical considerations for [RURAL MINISTRY CONTEXT]. How should [GEOGRAPHIC/CULTURAL CONTEXT] influence [MINISTERIAL APPROACH] to [RURAL COMMUNITY NEED]?

How to Use These Customizable Prompts

Customization Guidelines:

- **Be Specific**: Replace brackets with precise terms relevant to your research focus
- **Scale Appropriately**: Adjust complexity based on course level and assignment requirements
- **Context Matters**: Consider historical period, cultural setting, and target audience
- **Multiple Variables**: Some prompts can accommodate multiple topic insertions for comparative analysis

Example Customizations:

Original: "Compare [TRADITION 1] and [TRADITION 2] approaches to [THEOLOGICAL ISSUE]"

Customized: "Compare Lutheran and Reformed approaches to predestination" **Or**: "Compare Eastern Orthodox and Roman Catholic approaches to the Filioque controversy" **Or**: "Compare liberation theology and prosperity theology approaches to poverty"

Research Applications:

- **Term Papers**: Use single prompts as thesis foundations
- **Comparative Studies**: Combine multiple traditions or time periods
- **Case Studies**: Apply theological principles to specific situations
- **Thematic Analysis**: Trace concepts across different contexts
- **Contemporary Relevance**: Connect historical theology to current issues

75 Customizable AI Prompts for Philosophy and Religion Studies

75 Customizable AI Prompts for Philosophy and Religion Studies
Critical Inquiry and Analysis for Undergraduate Research

Philosophy of Religion (15 Prompts)

1. **Classical Arguments**: Critically evaluate the [ARGUMENT TYPE] argument for God's existence as presented by [PHILOSOPHER 1], [PHILOSOPHER 2], and [PHILOSOPHER 3]. What are the strongest objections and how do contemporary defenders respond to [SPECIFIC CRITICISM]?

2. **Problem of Evil Analysis**: Compare [THEODICY 1], [THEODICY 2], and [THEODICY 3] approaches to addressing the [LOGICAL/EVIDENTIAL] problem of [EVIL/SUFFERING]. Which approach most successfully preserves [DIVINE ATTRIBUTE] while accounting for [TYPE OF EVIL]?

3. **Religious Experience**: Analyze [PHILOSOPHER/THEOLOGIAN]'s criteria for authentic [RELIGIOUS EXPERIENCE TYPE]. How do [PHENOMENOLOGICAL/EPISTEMOLO

GICAL] critiques challenge the evidential value of [MYSTICAL/SPIRITUAL] experiences?

4. **Faith and Reason**: Examine the relationship between [FAITH] and [REASON] in [THINKER 1], [THINKER 2], and [CONTEMPORARY APPROACH]. What models best describe their interaction regarding [RELIGIOUS KNOWLEDGE/BELIEF JUSTIFICATION]?

5. **Divine Attributes**: Analyze the philosophical problems surrounding divine [ATTRIBUTE 1], [ATTRIBUTE 2], and [ATTRIBUTE 3]. How do different conceptions of God address these puzzles in [THEOLOGICAL TRADITION]?

6. **Religious Language**: Compare [PHILOSOPHICAL CRITIQUE] of religious language with responses from [PHILOSOPHICAL RESPONSE/SCHOOL]. What is the cognitive status of [THEOLOGICAL STATEMENTS/RELIGIOUS DISCOURSE] regarding [METAPHYSICAL CLAIMS]?

7. **Miracles and Natural Law**: Examine [PHILOSOPHER]'s argument against [MIRACLES/SUPERNATURAL EVENTS] and [CONTEMPORARY RESPONSES]. What constitutes adequate evidence for [MIRACULOUS EVENTS] in [RELIGIOUS CONTEXT]?

8. **Religious Diversity**: Analyze philosophical approaches to religious [PLURALISM/INCLUSIVISM/EXCLUSIVISM] regarding [TRUTH CLAIMS/SALVATION]. How should the fact of [RELIGIOUS DIVERSITY] affect [RELIGIOUS BELIEF/TRUTH CLAIMS]?
9. **Immortality and Personal Identity**: Examine philosophical arguments for [SURVIVAL AFTER DEATH/IMMORTALITY] in [PHILOSOPHICAL TRADITION]. What concept of [PERSONAL IDENTITY] best supports belief in [AFTERLIFE/CONTINUITY]?

10. **Divine Command Theory**: Analyze the relationship between [MORALITY] and [DIVINE COMMANDS] in [ETHICAL THEORY]. Does the [EUTHYPHRO DILEMMA/PHILOSOPHICAL OBJECTION] pose an insurmountable

challenge to [DIVINE COMMAND ETHICS]?

11. **Natural Theology**: Compare [CLASSICAL NATURAL THEOLOGY] with [CONTEMPORARY APPROACH] arguments for [RELIGIOUS BELIEF/GOD'S EXISTENCE]. What role should [PHILOSOPHY/REASON] play in supporting [RELIGIOUS BELIEF]?

12. **Religious Epistemology**: Examine [FOUNDATIONALIST/COHERENTIST] approaches to [RELIGIOUS KNOWLEDGE] in [PHILOSOPHICAL SCHOOL]. How should [RELIGIOUS BELIEFS] be justified philosophically regarding [KNOWLEDGE CLAIMS]?

13. **Temporal and Eternal**: Analyze competing philosophical models of divine [ETERNITY/TEMPORALITY] in [THEOLOGICAL TRADITION]. What implications do these have for [DIVINE-HUMAN INTERACTION/PRAYER]?

14. **Free Will and Divine Foreknowledge**: Examine the philosophical tension between [HUMAN FREEDOM] and divine [OMNISCIENCE/FOREKNOWLEDGE].

What solutions best preserve both concepts in [THEOLOGICAL FRAMEWORK]?

15. **Religious Realism vs. Anti-Realism**: Compare [REALIST] and [ANTI-REALIST] interpretations of [RELIGIOUS DISCOURSE/STATEMENTS]. Should [RELIGIOUS CLAIMS] be understood as [LITERALLY TRUE/METAPHORICALLY MEANINGFUL]?

Comparative Religion and Philosophy (15 Prompts)

16. **Eastern and Western Concepts**: Compare [EASTERN PHILOSOPHICAL CONCEPT] with [WESTERN PHILOSOPHICAL CONCEPT] regarding [PHILOSOPHICAL TOPIC]. What philosophical implications emerge from these different conceptions of [REALITY/SELF/CONSCIOUSNESS]?

17. **Monotheism vs. Non-Dualism**: Analyze the philosophical differences between [ABRAHAMIC TRADITION] monotheism and [EASTERN TRADITION] non-dualism. How do these worldviews address [ULTIMATE REALITY/DIVINE NATURE]?

18. **Religious Ethics Comparison**: Compare [VIRTUE ETHICS] in [RELIGIOUS TRADITION 1], [RELIGIOUS TRADITION 2], and [RELIGIOUS TRADITION 3]. What [UNIVERSAL/PARTICULAR] features characterize [RELIGIOUS MORAL SYSTEMS] regarding [ETHICAL ISSUE]?

19. **Concepts of Time**: Examine [LINEAR TIME CONCEPT] in [RELIGIOUS TRADITION 1] versus [CYCLICAL TIME CONCEPT] in [RELIGIOUS TRADITION 2]. What [METAPHYSICAL/ETHICAL] implications follow from these [TEMPORAL CONCEPTS] for [HUMAN EXISTENCE/SALVATION]?

20. **Mysticism Across Traditions**: Compare mystical experiences and philosophical interpretations in [TRADITION 1], [TRADITION 2], and [TRADITION 3] mysticism. What [COMMON FEATURES/SIGNIFICANT DIFFERENCES] emerge regarding [ULTIMATE REALITY/RELIGIOUS EXPERIENCE]?

21. **Religious Authority**: Analyze different sources of religious authority:

[AUTHORITY SOURCE 1], [AUTHORITY SOURCE 2], and [AUTHORITY SOURCE 3] across [RELIGIOUS TRADITIONS]. How do these interact and conflict regarding [RELIGIOUS KNOWLEDGE/PRACTICE]?

22. **Salvation and Liberation**: Compare [CHRISTIAN SOTERIOLOGICAL CONCEPT] with [BUDDHIST LIBERATION CONCEPT] and [HINDU MOKSHA CONCEPT]. What philosophical assumptions underlie different paths to [ULTIMATE FULFILLMENT/SPIRITUAL GOAL]?

23. **Prayer and Meditation**: Examine the philosophical foundations of [CONTEMPLATIVE PRACTICE] across [RELIGIOUS TRADITIONS]. What theories of [CONSCIOUSNESS/DIVINE INTERACTION] are presupposed in [SPIRITUAL PRACTICE]?

24. **Religious Community**: Compare [INDIVIDUALISTIC] and [COMMUNITARIAN] approaches to [RELIGIOUS LIFE] in [PHILOSOPHICAL TRADITION 1] and [PHILOSOPHICAL TRADITION 2]. How do concepts of

[SELF/COMMUNITY] affect [RELIGIOUS PRACTICE]?

25. **Sacred and Secular**: Analyze [PHILOSOPHER/SOCIOLOGIST]'s distinction between [SACRED/PROFANE] across [RELIGIOUS WORLDVIEWS]. How do various traditions understand the relationship between [ORDINARY REALITY] and [ULTIMATE REALITY]?

26. **Religious Art and Symbolism**: Examine the philosophical significance of [RELIGIOUS AESTHETICS] in [RELIGIOUS TRADITION 1], [RELIGIOUS TRADITION 2], and [RELIGIOUS TRADITION 3]. How does each tradition understand the relationship between [BEAUTY] and [TRUTH/DIVINE]?

27. **Pilgrimage and Sacred Space**: Compare philosophical concepts of [SACRED GEOGRAPHY] in [RELIGIOUS TRADITION 1] and [RELIGIOUS TRADITION 2]. What [METAPHYSICAL ASSUMPTIONS] underlie beliefs about [HOLY PLACES/SACRED SPACE]?

28. **Religious Fundamentalism**: Analyze the philosophical characteristics of

[FUNDAMENTALIST MOVEMENT] across [RELIGIOUS TRADITIONS]. What common [EPISTEMOLOGICAL/HERMENEUTICAL] features emerge regarding [RELIGIOUS AUTHORITY/INTERPRETATION]?

29. **Religion and Politics**: Compare [THEOCRATIC], [SECULAR], and [DHARMIC] approaches to the relationship between [RELIGIOUS AUTHORITY] and [POLITICAL AUTHORITY]. What philosophical principles guide [CHURCH-STATE RELATIONS] in [CULTURAL CONTEXT]?

30. **Interfaith Philosophy**: Examine philosophical approaches to [RELIGIOUS DIALOGUE] between [RELIGION 1] and [RELIGION 2]. What [EPISTEMOLOGICAL/METAPHYSICAL] assumptions facilitate or hinder [INTERFAITH ENGAGEMENT] regarding [SHARED CONCERN]?

Ancient Philosophy and Religion (15 Prompts)

31. **Platonic Religion**: Analyze [PLATO]'s concept of [THE GOOD/DIVINE] and its influence on [RELIGIOUS THOUGHT/THEOLOGICAL TRADITION]. How does [PLATONIC METAPHYSICS] shape [THEOLOGICAL SPECULATION] about [DIVINE NATURE]?

32. **Aristotelian Theology**: Examine [ARISTOTLE]'s [PRIME MOVER ARGUMENT] and its reception in [CHRISTIAN/ISLAMIC/JEWISH] thought. What philosophical problems does the [UNMOVED MOVER] address regarding [CAUSATION/DIVINE ACTION]?

33. **Stoicism and Providence**: Analyze [STOIC] concepts of [FATE/PROVIDENCE/VIRTUE] in [STOIC PHILOSOPHER]. How do [STOIC PHILOSOPHICAL PRINCIPLES] compare with [RELIGIOUS CONCEPTS] of [DIVINE GOVERNANCE]?

34. **Epicureanism vs. Religion**: Examine [EPICUREAN] critiques of [POPULAR RELIGION] and their philosophical

foundations in [EPICUREAN MATERIALISM]. What challenges does [EPICUREAN PHILOSOPHY] pose to [RELIGIOUS WORLDVIEWS] regarding [DIVINE INTERVENTION/AFTERLIFE]?

35. **Neo-Platonism**: Analyze [PLOTINUS]'s [EMANATIONIST METAPHYSICS] and its influence on [CHRISTIAN/ISLAMIC/JEWISH] mysticism. How does [NEO-PLATONIC PHILOSOPHY] shape religious thought about [DIVINE TRANSCENDENCE/UNION]?

36. **Pre-Socratic Cosmology**: Compare early [GREEK COSMOLOGICAL SPECULATIONS] with [ANCIENT NEAR EASTERN CREATION MYTHS]. What philosophical innovations emerge in [GREEK THOUGHT] regarding [NATURAL EXPLANATION/DIVINE CAUSATION]?

37. **Hellenistic Mystery Religions**: Examine the philosophical elements in [MYSTERY CULT] and their relationship to [POPULAR PHILOSOPHY]. How do [RITUAL PRACTICE] and [PHILOSOPHICAL REFLECTION] interact in [RELIGIOUS EXPERIENCE]?

38. **Ancient Skepticism**: Analyze [ACADEMIC/PYRRHONIAN] skepticism regarding [RELIGIOUS KNOWLEDGE] in [SKEPTICAL PHILOSOPHER]. What challenges do [ANCIENT SKEPTICAL ARGUMENTS] pose to [RELIGIOUS EPISTEMOLOGY]?

39. **Patristic Philosophy**: Examine how early Christian thinkers like [CHURCH FATHER] integrated [GREEK PHILOSOPHY] with [CHRISTIAN THEOLOGY]. What synthesis emerges regarding [PHILOSOPHICAL/THEOLOGICAL ISSUE]?

40. **Ancient Ethics and Religion**: Compare [PLATONIC/ARISTOTELIAN/STOIC] virtue ethics with [RELIGIOUS MORAL TEACHING]. What philosophical foundations support [ETHICAL SYSTEMS] in [ANCIENT CONTEXT]?

41. **Ancient Political Philosophy**: Analyze [PLATO/ARISTOTLE]'s views on [RELIGION'S ROLE] in the [IDEAL STATE]. How should [PHILOSOPHY] and [RELIGION] interact in [POLITICAL LIFE]?

42. **Gnosticism**: Examine the philosophical foundations of [GNOSTIC DUALISM] and its critique by [ORTHODOX CHRISTIANITY]. What [METAPHYSICAL ISSUES] are at stake regarding [MATERIAL/SPIRITUAL REALITY]?

43. **Ancient Atomism**: Analyze [DEMOCRITEAN/EPICUREAN] atomism as [ALTERNATIVE] to [RELIGIOUS EXPLANATION] of [REALITY/CAUSATION]. What philosophical arguments support [MATERIALIST WORLDVIEW] against [RELIGIOUS COSMOLOGY]?

44. **Pythagorean Philosophy**: Examine the [RELIGIOUS/PHILOSOPHICAL] elements in [PYTHAGOREANISM]. How do [MATHEMATICAL] and [MYSTICAL] concerns interact in [PYTHAGOREAN THOUGHT]?

45. **Hermeticism**: Analyze the philosophical theology of [HERMETIC TEXTS] and their influence on [LATER THOUGHT]. What synthesis of [GREEK PHILOSOPHY] and [RELIGIOUS WISDOM] emerges regarding [DIVINE KNOWLEDGE]?

Medieval Philosophy and Religion (15 Prompts)

46. **Islamic Philosophy**: Compare [AL-FARABI/AVICENNA/AVERROES] on the relationship between [PHILOSOPHY] and [REVEALED RELIGION]. How do they resolve apparent conflicts regarding [REASON/REVELATION]?

47. **Jewish Medieval Thought**: Analyze [MAIMONIDES]' [GUIDE FOR THE PERPLEXED] and its attempt to harmonize [ARISTOTELIAN PHILOSOPHY] with [JEWISH THEOLOGY]. What philosophical method emerges for [RELIGIOUS RATIONALISM]?

48. **Christian Scholasticism**: Examine the philosophical method of [THOMAS AQUINAS] in addressing [THEOLOGICAL QUESTIONS]. How does his synthesis of [ARISTOTLE] and [CHRISTIANITY] work regarding [FAITH/REASON]?

49. **Mystical Philosophy**: Compare the philosophical foundations of [CHRISTIAN MYSTICISM], [ISLAMIC SUFISM], and [JEWISH KABBALAH] during the [MEDIEVAL PERIOD]. What

[PHILOSOPHICAL FRAMEWORK] supports [MYSTICAL EXPERIENCE]?

50. **Faith and Reason Controversy**: Analyze the medieval debate over the relationship between [PHILOSOPHICAL TRUTH] and [RELIGIOUS TRUTH], including the role of [DOUBLE TRUTH THEORY] in [AVERROISM/SCHOLASTICISM].

51. **Islamic Occasionalism**: Examine [AL-GHAZALI]'s critique of [ARISTOTELIAN PHILOSOPHY] and his [OCCASIONALIST ALTERNATIVE]. What [THEOLOGICAL CONCERNS] motivate his philosophical position regarding [CAUSATION/DIVINE ACTION]?

52. **Medieval Nominalism**: Analyze [OCKHAM]'s [NOMINALISM] and its implications for [THEOLOGY]. How does his philosophical position affect [RELIGIOUS LANGUAGE] and [THEOLOGICAL KNOWLEDGE]?

53. **Averroist Controversy**: Examine the reception of [AVERROES] in [MEDIEVAL CHRISTIAN THOUGHT] and the philosophical issues surrounding [SEPARATE/PERSONAL

IMMORTALITY] in [ARISTOTELIAN PSYCHOLOGY].

54. **Medieval Political Philosophy**: Compare [ISLAMIC], [JEWISH], and [CHRISTIAN] views on the relationship between [RELIGIOUS AUTHORITY] and [POLITICAL AUTHORITY] in [MEDIEVAL CONTEXT].

55. **Sufism and Philosophy**: Analyze the philosophical dimensions of [SUFI THOUGHT], particularly in [IBN SINA/SUHRAWARDI]. How do [MYSTICAL EXPERIENCE] and [RATIONAL REFLECTION] interact in [ISLAMIC PHILOSOPHY]?

56. **Medieval Cosmology**: Examine [MEDIEVAL CHRISTIAN/ISLAMIC/JEWISH] philosophical approaches to understanding [STRUCTURE OF REALITY] and [GOD'S RELATIONSHIP] to [CREATION].

57. **Medieval Epistemology**: Compare medieval approaches to [RELIGIOUS KNOWLEDGE], including the roles of [REASON], [REVELATION], and [MYSTICAL EXPERIENCE] in

[SCHOLASTIC/ISLAMIC/JEWISH] traditions.

58. **Natural Law Theory**: Analyze the development of [NATURAL LAW] concepts in [MEDIEVAL PHILOSOPHY] and their relationship to [DIVINE COMMAND] and [ETERNAL LAW] in [THOMISTIC THOUGHT].

59. **Medieval Logic**: Examine how developments in [MEDIEVAL LOGIC] affect [THEOLOGICAL REASONING] and [RELIGIOUS LANGUAGE] in [OCKHAM/SCOTUS].

60. **University and Religion**: Analyze the philosophical implications of [MEDIEVAL UNIVERSITIES] for the relationship between [SECULAR LEARNING] and [RELIGIOUS AUTHORITY] in [SCHOLASTIC CONTEXT].

Modern Philosophy and Religion (15 Prompts)

61. **Cartesian Dualism**: Analyze [DESCARTES]' [MIND-BODY DUALISM] and its implications for understanding [THE SOUL/IMMORTALITY/RELIGIOUS EXPERIENCE]. What problems does this create for [PERSONAL IDENTITY/DIVINE INTERACTION]?

62. **Spinoza's Pantheism**: Examine [SPINOZA]'s [PHILOSOPHICAL THEOLOGY] and its challenge to traditional [THEISM/DIVINE TRANSCENDENCE]. How does his [GEOMETRIC METHOD] affect religious concepts of [GOD/PRAYER/PROVIDENCE]?

63. **Empiricism and Religion**: Compare [LOCKE]'s [RATIONAL CHRISTIANITY] with [HUME]'s [SKEPTICAL CRITIQUE] of [RELIGION/MIRACLES]. What role does [EMPIRICISM] play in evaluating [RELIGIOUS CLAIMS/SUPERNATURAL EVENTS]?

64. **Kant's Critical Philosophy**: Analyze [KANT]'s critique of traditional arguments for [GOD'S EXISTENCE] and his [MORAL ARGUMENT]. How does [TRANSCENDENTAL IDEALISM] affect [RELIGIOUS KNOWLEDGE/METAPHYSICAL CLAIMS]?

65. **German Idealism**: Examine [HEGEL]'s philosophical treatment of [RELIGION] as a stage in the development of [ABSOLUTE SPIRIT]. What is religion's relationship to [PHILOSOPHY/RATIONAL KNOWLEDGE] in [HEGELIAN SYSTEM]?

66. **Enlightenment Deism**: Analyze the philosophical foundations of [DEISTIC RELIGION] and its critique of [REVEALED RELIGION]. What [RATIONAL RELIGION] emerges from [ENLIGHTENMENT THOUGHT] regarding [NATURAL THEOLOGY]?

67. **Pascal's Wager**: Examine [PASCAL]'s [PROBABILISTIC ARGUMENT] for [RELIGIOUS BELIEF] and contemporary philosophical critiques. What [DECISION-THEORETIC ISSUES] arise regarding

[BELIEF FORMATION/RELIGIOUS COMMITMENT]?

68. **Berkeley's Immaterialism**: Analyze [BERKELEY]'s [PHILOSOPHICAL IDEALISM] and its theological implications. How does ["ESSE EST PERCIPI"] support [RELIGIOUS WORLDVIEW/DIVINE PRESENCE]?

69. **Kierkegaard's Existentialism**: Examine [KIERKEGAARD]'s critique of [HEGELIAN PHILOSOPHY] and his emphasis on [FAITH] as [PASSIONATE COMMITMENT]. What view of [RELIGIOUS TRUTH/AUTHENTIC EXISTENCE] emerges from [EXISTENTIALIST APPROACH]?

70. **Nietzsche's Critique**: Analyze [NIETZSCHE]'s ["DEATH OF GOD"] thesis and its implications for [MORALITY/MEANING]. What philosophical challenges does he pose to [RELIGIOUS BELIEF/TRADITIONAL VALUES]?

71. **Schleiermacher's Theology**: Examine [SCHLEIERMACHER]'s philosophical approach to [RELIGION] through [FEELING/ABSOLUTE DEPENDENCE].

How does [ROMANTICISM] reshape [RELIGIOUS THOUGHT/THEOLOGICAL METHOD]?

72. **Mill's Utilitarianism**: Analyze [JOHN STUART MILL]'s approach to [RELIGION] and its compatibility with [UTILITARIAN ETHICS]. What role should [RELIGION] play in [MORAL CALCULATION/SOCIAL WELFARE]?

73. **Feuerbach's Projection Theory**: Examine [FEUERBACH]'s [ANTHROPOLOGICAL CRITIQUE] of [RELIGION] and its influence on [LATER PHILOSOPHY]. What philosophical method underlies his analysis of [RELIGIOUS BELIEF/GOD-CONCEPTS]?

74. **Schopenhauer's Pessimism**: Analyze [SCHOPENHAUER]'s philosophical system and its relationship to [EASTERN RELIGIOUS THOUGHT], particularly [BUDDHISM]. What synthesis emerges regarding [SUFFERING/LIBERATION/WILL]?

75. **Religious Romanticism**: Examine the [ROMANTIC MOVEMENT]'s rehabilitation of [RELIGIOUS EXPERIENCE] and its philosophical

foundations. How does [ROMANTICISM] challenge [ENLIGHTENMENT RATIONALISM] regarding [RELIGIOUS KNOWLEDGE/SPIRITUAL EXPERIENCE]?

How to Use These Customizable Prompts for Advanced Study

Customization Strategies:

Single Focus: Replace all brackets with one consistent theme

- Example: "Analyze [KANT]'s critique of traditional arguments for [GOD'S EXISTENCE]" becomes "Analyze Kant's critique of traditional arguments for God's existence"

Comparative Analysis: Use contrasting terms in brackets

- Example: "Compare [RATIONAL CHRISTIANITY] with [SKEPTICAL CRITIQUE]" becomes "Compare Locke's rational Christianity with Hume's skeptical critique"

Contemporary Application: Insert modern issues or thinkers

- Example: "What philosophical challenges does [NIETZSCHE] pose to [RELIGIOUS BELIEF]" becomes "What philosophical challenges does Nietzsche pose to contemporary evangelical Christianity"

Cross-Cultural Focus: Emphasize different religious traditions

- Example: "Compare [EASTERN PHILOSOPHICAL CONCEPT] with [WESTERN PHILOSOPHICAL CONCEPT]" becomes "Compare Buddhist no-self doctrine with Cartesian personal identity"

Advanced Research Applications:

Thesis Development: Use prompts to generate original research questions that combine multiple philosophical traditions, historical periods, or contemporary issues.

Comprehensive Examinations: Practice synthesizing major philosophical movements, key arguments, and their religious implications across different time periods.

Graduate Preparation: Build familiarity with primary sources, secondary scholarship, and contemporary debates in philosophy of religion.

Interdisciplinary Research: Connect philosophical analysis with insights from psychology, anthropology, cognitive science, and other relevant fields.

Critical Thinking Enhancement:

Argument Analysis: Practice identifying premises, conclusions, and logical structures in complex philosophical arguments about religious topics.

Historical Contextualization: Understand how social, cultural, and intellectual contexts shape philosophical and religious developments.
Comparative Methodology: Develop skills in fair and rigorous comparison across different philosophical and religious traditions.

Contemporary Relevance: Apply historical philosophical insights to current debates in religion, ethics, politics, and culture.

90 AI Prompts for Philosophy and Religion Studies

Critical Inquiry and Analysis for Undergraduate Research

90 AI Prompts for Philosophy and Religion Studies

Critical Inquiry and Analysis for Undergraduate Research

A collection of 90 additional AI prompts specifically designed for Philosophy and Religion studies. These prompts are organized into six sophisticated categories:

1. **Philosophy of Religion** (15 prompts) - Core philosophical questions about religious belief, experience, and knowledge

2. **Comparative Religion and Philosophy** (15 prompts) - Cross-cultural philosophical analysis of religious traditions

3. **Ancient Philosophy and Religion** (15 prompts) - Classical foundations of religious philosophical thought

4. **Medieval Philosophy and Religion** (15 prompts) - Scholastic synthesis and interfaith philosophical dialogue

5. **Modern Philosophy and Religion** (15 prompts) - Enlightenment through 19th-century developments

6. **Contemporary Issues** (15 prompts) - Current debates and emerging questions in philosophy of religion

These prompts are designed to:

- **Develop Critical Thinking**: Each prompt requires analysis, evaluation, and synthesis of complex philosophical arguments
- **Bridge Historical and Contemporary Issues**: Connect classical philosophical problems with current debates
- **Encourage Comparative Analysis**: Examine how different philosophical and religious traditions approach similar questions
- **Foster Original Research**: Provide starting points for independent scholarly investigation
- **Build Philosophical Skills**: Practice argumentation, logical analysis, and conceptual clarification

The prompts are suitable for:

- **Upper-level undergraduate courses** in philosophy of religion, comparative religion, and related fields

- **Independent study projects** and honors theses

- **Graduate school preparation** in philosophy, religious studies, or theology

- **Seminar discussions** requiring sophisticated philosophical analysis

Each prompt is crafted to move beyond descriptive summary toward critical analysis, philosophical argumentation, and original insight—essential skills for advanced study in philosophy and religion.

90 AI Prompts for Philosophy and Religion Studies

Critical Inquiry and Analysis for Undergraduate Research

Philosophy of Religion (15 Prompts)

1. **Classical Arguments for God's Existence**: Critically evaluate the ontological argument as presented by Anselm, Descartes, and Plantinga. What are the strongest objections and how do contemporary defenders respond?

2. **Problem of Evil Analysis**: Compare Augustinian, Irenaean, and process theodicies in addressing the logical and evidential problems of evil. Which approach most successfully preserves divine attributes?

3. **Religious Experience**: Analyze William James' criteria for authentic religious experience. How do phenomenological and epistemological critiques challenge the evidential value of mystical experiences?

4. **Faith and Reason**: Examine the relationship between faith and reason in Aquinas, Kierkegaard, and contemporary

reformed epistemology. What models best describe their interaction?

5. **Divine Attributes**: Analyze the philosophical problems surrounding divine omniscience, omnipotence, and perfect goodness. How do different conceptions of God address these puzzles?

6. **Religious Language**: Compare logical positivist critiques of religious language with responses from Wittgensteinian philosophy of religion. What is the cognitive status of theological statements?

7. **Miracles and Natural Law**: Examine Hume's argument against miracles and contemporary responses. What constitutes adequate evidence for miraculous events?

8. **Religious Diversity**: Analyze philosophical approaches to religious pluralism, inclusivism, and exclusivism. How should the fact of religious diversity affect religious belief?
9. **Immortality and Personal Identity**: Examine philosophical arguments for survival after death. What concept of personal identity best supports belief in an afterlife?

10. **Divine Command Theory**: Analyze the relationship between morality and divine commands. Does the Euthyphro dilemma pose an insurmountable challenge to divine command ethics?

11. **Natural Theology**: Compare classical natural theology with contemporary intelligent design arguments. What role should philosophy play in supporting religious belief?

12. **Religious Epistemology**: Examine foundationalist versus coherentist approaches to religious knowledge. How should religious beliefs be justified philosophically?

13. **Temporal and Eternal**: Analyze competing philosophical models of divine eternity versus divine temporality. What implications do these have for divine-human interaction?

14. **Free Will and Divine Foreknowledge**: Examine the philosophical tension between human freedom and divine omniscience. What solutions best preserve both concepts?

15. **Religious Realism vs. Anti-Realism**: Compare realist and anti-realist

interpretations of religious discourse. Should religious statements be understood as literally true or metaphorically meaningful?

Comparative Religion and Philosophy (15 Prompts)

16. **Eastern and Western Concepts of Self**: Compare Buddhist no-self doctrine with Cartesian dualism and contemporary theories of personal identity. What philosophical implications emerge from these different conceptions?

17. **Monotheism vs. Non-Dualism**: Analyze the philosophical differences between Abrahamic monotheism and Advaita Vedanta non-dualism. How do these worldviews address ultimate reality?
18. **Religious Ethics Comparison**: Compare virtue ethics in Aristotelian, Christian, Buddhist, and Islamic traditions. What universal and particular features characterize religious moral systems?

19. **Concepts of Time**: Examine linear time in Judeo-Christian thought versus cyclical time in Hindu and Buddhist philosophy. What metaphysical and ethical implications follow from these temporal concepts?

20. **Mysticism Across Traditions**: Compare mystical experiences and philosophical interpretations in Christian, Islamic (Sufi), and Hindu traditions. What common features and significant differences emerge?

21. **Religious Authority**: Analyze different sources of religious authority: scripture, tradition, reason, and experience across various religious philosophies. How do these interact and conflict?

22. **Salvation and Liberation**: Compare Christian soteriological concepts with Buddhist liberation and Hindu moksha. What philosophical assumptions underlie different paths to ultimate fulfillment?

23. **Prayer and Meditation**: Examine the philosophical foundations of contemplative practices across religious traditions. What theories of consciousness and divine interaction are presupposed?

24. **Religious Community**: Compare individualistic and communitarian approaches to religious life in different philosophical traditions. How do concepts of self-affect religious practice?

25. **Sacred and Secular**: Analyze Durkheim's distinction between sacred and profane across different religious worldviews. How do various traditions understand the relationship between ordinary and ultimate reality?

26. **Religious Art and Symbolism**: Examine the philosophical significance of religious aesthetics in Islamic, Christian, and Buddhist traditions. How does each tradition understand the relationship between beauty and truth?

27. **Pilgrimage and Sacred Space**: Compare philosophical concepts of sacred geography in different religious traditions. What metaphysical assumptions underlie beliefs about holy places?
28. **Religious Fundamentalism**: Analyze the philosophical characteristics of fundamentalist movements across different religions. What common epistemological and hermeneutical features emerge?

29. **Religion and Politics**: Compare theocratic, secular, and dharmic approaches to the relationship between religious and political authority. What philosophical principles guide church-state relations?

30. **Interfaith Philosophy**: Examine philosophical approaches to religious dialogue and mutual understanding. What epistemological and metaphysical assumptions facilitate or hinder interfaith engagement?

Ancient Philosophy and Religion (15 Prompts)

31. **Platonic Religion**: Analyze Plato's concept of the Good and its influence on later religious thought. How does Platonic metaphysics shape theological speculation?

32. **Aristotelian Theology**: Examine Aristotle's Prime Mover argument and its reception in Christian, Islamic, and Jewish thought. What philosophical problems does the Unmoved Mover address?

33. **Stoicism and Providence**: Analyze Stoic concepts of fate, providence, and virtue. How do Stoic philosophical principles compare with religious concepts of divine governance?

34. **Epicureanism vs. Religion**: Examine Epicurean critiques of popular religion and their philosophical foundations. What

challenges does Epicurean materialism pose to religious worldviews?

35. **Neo-Platonism**: Analyze Plotinus's emanationist metaphysics and its influence on Christian, Islamic, and Jewish mysticism. How does Neo-Platonic philosophy shape religious thought?

36. **Pre-Socratic Cosmology**: Compare early Greek cosmological speculations with ancient Near Eastern creation myths. What philosophical innovations emerge in Greek thought?

37. **Hellenistic Mystery Religions**: Examine the philosophical elements in mystery cults and their relationship to popular philosophy. How do ritual and philosophical reflection interact?

38. **Ancient Skepticism**: Analyze Academic and Pyrrhonian skepticism regarding religious knowledge. What challenges do ancient skeptical arguments pose to religious epistemology?

39. **Patristic Philosophy**: Examine how early Christian thinkers like Justin Martyr and Clement of Alexandria integrated Greek philosophy with Christian theology. What synthesis emerges?

40. **Ancient Ethics and Religion**: Compare Platonic, Aristotelian, and Stoic virtue ethics with religious moral teachings. What philosophical foundations support ethical systems?

41. **Ancient Political Philosophy**: Analyze Plato's and Aristotle's views on religion's role in the ideal state. How should philosophy and religion interact in political life?

42. **Gnosticism**: Examine the philosophical foundations of Gnostic dualism and its critique by orthodox Christianity. What metaphysical issues are at stake?

43. **Ancient Atomism**: Analyze Democritean and Epicurean atomism as alternatives to religious explanations of reality. What philosophical arguments support materialist worldviews?

44. **Pythagorean Philosophy**: Examine the religious and philosophical elements in Pythagoreanism. How do mathematical and mystical concerns interact?

45. **Hermeticism**: Analyze the philosophical theology of Hermetic texts and their influence on later thought. What synthesis of

Greek philosophy and religious wisdom emerges?

Medieval Philosophy and Religion (15 Prompts)

46. **Islamic Philosophy**: Compare Al-Farabi, Avicenna, and Averroes on the relationship between philosophy and revealed religion. How do they resolve apparent conflicts?

47. **Jewish Medieval Thought**: Analyze Maimonides' Guide for the Perplexed and its attempt to harmonize Aristotelian philosophy with Jewish theology. What philosophical method emerges?

48. **Christian Scholasticism**: Examine the philosophical method of Thomas Aquinas in addressing theological questions. How does his synthesis of Aristotle and Christianity work?

49. **Mystical Philosophy**: Compare the philosophical foundations of Christian (Eckhart), Islamic (Ibn Arabi), and Jewish (Kabbalah) mysticism during the medieval period.

50. **Faith and Reason Controversy**: Analyze the medieval debate over the relationship

between philosophical and religious truth, including the role of double truth theories.

51. **Islamic Occasionalism**: Examine Al-Ghazali's critique of Aristotelian philosophy and his occasionalist alternative. What theological concerns motivate his philosophical position?

52. **Medieval Nominalism**: Analyze Ockham's nominalism and its implications for theology. How does his philosophical position affect religious language and knowledge?

53. **Averroist Controversy**: Examine the reception of Averroes in medieval Christian thought and the philosophical issues surrounding separate and personal immortality.

54. **Medieval Political Philosophy**: Compare Islamic (Al-Farabi), Jewish (Maimonides), and Christian (Aquinas) views on the relationship between religious and political authority.

55. **Sufism and Philosophy**: Analyze the philosophical dimensions of Sufi thought, particularly in figures like Ibn Sina and

Suhrawardi. How do mystical experience and rational reflection interact?

56. **Medieval Cosmology**: Examine medieval Christian, Islamic, and Jewish philosophical approaches to understanding the structure of reality and God's relationship to creation.

57. **Medieval Epistemology**: Compare medieval approaches to religious knowledge, including the roles of reason, revelation, and mystical experience in different traditions.

58. **Natural Law Theory**: Analyze the development of natural law concepts in medieval philosophy and their relationship to divine command and eternal law.

59. **Medieval Logic**: Examine how developments in medieval logic (particularly in the work of Ockham and Scotus) affect theological reasoning and religious language.

60. **University and Religion**: Analyze the philosophical implications of the rise of medieval universities for the relationship between secular learning and religious authority.

Modern Philosophy and Religion (15 Prompts)

61. **Cartesian Dualism**: Analyze Descartes' mind-body dualism and its implications for understanding the soul, immortality, and religious experience. What problems does this create?

62. **Spinoza's Pantheism**: Examine Spinoza's philosophical theology and its challenge to traditional theism. How does his geometric method affect religious concepts?

63. **Empiricism and Religion**: Compare Locke's rational Christianity with Hume's skeptical critique of religion. What role does empiricism play in evaluating religious claims?
64. **Kant's Critical Philosophy**: Analyze Kant's critique of traditional arguments for God's existence and his moral argument. How does transcendental idealism affect religious knowledge?

65. **German Idealism**: Examine Hegel's philosophical treatment of religion as a stage in the development of absolute spirit. What is religion's relationship to philosophy?

66. **Enlightenment Deism**: Analyze the philosophical foundations of Deistic religion and its critique of revealed religion. What rational religion emerges from Enlightenment thought?

67. **Pascal's Wager**: Examine Pascal's probabilistic argument for religious belief and contemporary philosophical critiques. What decision-theoretic issues arise?

68. **Berkeley's Immaterialism**: Analyze Berkeley's philosophical idealism and its theological implications. How does "esse est percipi" support religious worldviews?

69. **Kierkegaard's Existentialism**: Examine Kierkegaard's critique of Hegelian philosophy and his emphasis on faith as passionate commitment. What view of religious truth emerges?

70. **Nietzsche's Critique**: Analyze Nietzsche's "death of God" thesis and its implications for morality and meaning. What philosophical challenges does he pose to religious belief?

71. **Schleiermacher's Theology**: Examine Schleiermacher's philosophical approach to religion through feeling and absolute

dependence. How does romanticism reshape religious thought?

72. **Mill's Utilitarianism**: Analyze John Stuart Mill's approach to religion and its compatibility with utilitarian ethics. What role should religion play in moral calculation?

73. **Feuerbach's Projection Theory**: Examine Feuerbach's anthropological critique of religion and its influence on later philosophy. What philosophical method underlies his analysis?
74. **Schopenhauer's Pessimism**: Analyze Schopenhauer's philosophical system and its relationship to Eastern religious thought, particularly Buddhism. What synthesis emerges?

75. **Religious Romanticism**: Examine the romantic movement's rehabilitation of religious experience and its philosophical foundations. How does romanticism challenge Enlightenment rationalism?

Contemporary Issues in Philosophy and Religion (15 Prompts)

76. **Postmodern Religion**: Analyze postmodern critiques of traditional religious

authority and truth claims. How do Derrida, Foucault, and others challenge religious epistemology?

77. **Feminist Philosophy of Religion**: Examine feminist critiques of traditional religious concepts and practices. What alternative philosophical approaches to religion emerge?

78. **Science and Religion**: Analyze contemporary philosophical approaches to the relationship between scientific and religious worldviews. What models of interaction are proposed?

79. **Process Philosophy**: Examine Whitehead's process philosophy and its theological applications in process theology. What alternative concept of divinity emerges?

80. **Analytic Philosophy of Religion**: Analyze how analytic philosophical methods have transformed contemporary philosophy of religion. What new questions and approaches emerge?

81. **Continental Philosophy**: Examine how continental philosophical traditions (phenomenology, hermeneutics,

existentialism) approach religious questions differently from analytic philosophy.

82. **Religious Naturalism**: Analyze philosophical attempts to develop naturalistic approaches to religion. How do thinkers like Dewey and Santayana reconceive religious experience?
83. **Cognitive Science of Religion**: Examine how developments in cognitive science challenge or support traditional philosophical approaches to religious belief and experience.

84. **Global Philosophy**: Analyze how increased engagement with non-Western philosophical traditions affects contemporary philosophy of religion in academic settings.

85. **Environmental Philosophy**: Examine how environmental concerns reshape contemporary religious philosophy. What eco-theological philosophical positions emerge?

86. **Technology and Religion**: Analyze the philosophical implications of digital technology for religious practice, community, and authority. What new questions arise?

87. **Medical Ethics and Religion**: Examine how advances in medical technology create new philosophical questions about the intersection of religious and ethical reasoning.

88. **Political Theology**: Analyze contemporary philosophical approaches to the relationship between religious belief and political engagement in pluralistic societies.

89. **Religious Epistemology**: Examine contemporary debates in religious epistemology, including reformed epistemology, religious experientialism, and evidential arguments.

90. **Philosophy of Religious Diversity**: Analyze contemporary philosophical approaches to religious pluralism and the challenge of conflicting truth claims among world religions.

How to Use These Prompts for Advanced Study

For Research Papers:

- Develop original thesis statements from comparative analyses
- Use prompts to identify gaps in current scholarship
- Create interdisciplinary approaches combining philosophical and religious studies methods

For Seminar Discussions:

- Generate sophisticated philosophical dialogue on religious questions
- Practice defending philosophical positions with rigorous argumentation
- Explore the intersection of historical and contemporary issues

For Comprehensive Examinations:

- Master major figures, movements, and concepts in philosophy of religion
- Practice synthesizing different philosophical traditions and approaches
- Develop ability to analyze complex philosophical arguments

For Graduate School Preparation:

- Build familiarity with major debates in contemporary philosophy of religion

- Develop skills in philosophical analysis and argumentation
- Create foundation for specialized research in philosophy and religious studies

Advanced Research Strategies:

- Engage with both historical sources and contemporary scholarship
- Apply rigorous philosophical methods to religious questions
- Develop original insights through comparative and critical analysis
- Consider interdisciplinary approaches combining philosophy, theology, anthropology, and other fields

Critical Thinking Development:

- Question assumptions underlying religious and philosophical positions
- Analyze the logic and coherence of complex arguments
- Evaluate evidence and reasoning in religious and philosophical contexts
- Develop skills in constructive criticism and charitable interpretation

75 Customizable AI Prompts for Master of Divinity and Theological Studies

75 Customizable AI Prompts for Master of Divinity and Theological Studies

75 customizable AI prompts specifically designed for the Master of Divinity (M.Div.) and Theological Studies programs. These prompts reflect the advanced academic rigor and practical ministry preparation expected at the graduate seminary level.

Key Features of These Graduate-Level Prompts:

1. **Advanced Academic Rigor**: Each prompt requires engagement with primary sources, sophisticated theological methodology, and graduate-level analysis appropriate for seminary education.

2. **Ministerial Preparation Focus**: The prompts bridge academic theology with practical ministry concerns, reflecting the M.Div.'s professional degree orientation.

3. **Denominational Flexibility**: Students can insert their specific confessional tradition

while engaging ecumenically with broader theological resources.

4. **Interdisciplinary Integration**: Prompts encourage integration of theology with biblical studies, church history, ethics, and pastoral care.

5. **Contemporary Application**: Each prompt connects classical theological resources with current church and world challenges.

The Five Categories Include:

- **Systematic Theology and Doctrine Development** (15 prompts): For constructive theological work and doctrinal analysis

- **Biblical Exegesis and Hermeneutics** (15 prompts): For advanced biblical interpretation and theological exegesis

- **Historical Theology and Church History** (15 prompts): For engagement with theological tradition and historical development

- **Pastoral Theology and Practical Ministry** (15 prompts): For integrating theology with ministry practice

- **Ethics and Moral Theology** (15 prompts): For applying theological principles to contemporary ethical challenges

Graduate-Level Distinctions:

These prompts differ from undergraduate versions by:

- Requiring primary source engagement rather than secondary summaries
- Expecting original theological contribution beyond descriptive analysis
- Integrating multiple theological disciplines within single prompts
- Addressing practical ministry preparation and pastoral application
- Assuming denominational/confessional context while encouraging ecumenical engagement
- Supporting thesis-level research and comprehensive examination preparation

Practical Applications:

- Master's thesis development and research
- Comprehensive examination preparation
- Professional ministry skill development
- Academic conference presentation preparation
- Denominational position paper development

- Advanced coursework in systematic theology, biblical studies, church history, and practical theology

These prompts can be scaled from seminar papers to major research projects, providing flexibility for different course requirements while maintaining the academic and pastoral integration essential to quality theological education.

75 Customizable AI Prompts for Master of Divinity and Theological Studies

Instructions: Replace the bracketed placeholders [TOPIC], [TRADITION], [THEOLOGIAN], [DOCTRINE], etc. with your specific research focus, denominational context, or theological interest. These prompts are designed for graduate-level analysis and ministerial preparation.

Systematic Theology and Doctrine Development (15 prompts)

1. **Constructive Doctrinal Synthesis**: Develop a contemporary systematic treatment of [DOCTRINE] that engages [CLASSICAL THEOLOGIAN], [CONTEMPORARY

THEOLOGIAN], and [PHILOSOPHICAL CHALLENGE]. How might this doctrine be reformulated to address [MODERN PASTORAL/THEOLOGICAL CONCERN] while maintaining [CONFESSIONAL/TRADITIONAL BOUNDARIES]?

2. **Theological Method Integration**: Construct a theological methodology that integrates [THEOLOGICAL SOURCE 1], [THEOLOGICAL SOURCE 2], and [THEOLOGICAL SOURCE 3] for addressing [CONTEMPORARY THEOLOGICAL QUESTION]. What hermeneutical principles govern the relationship between these sources when they appear to conflict regarding [SPECIFIC ISSUE]?

3. **Dogmatic Development Analysis**: Trace the development of [SPECIFIC DOCTRINE] from [EARLY PERIOD] through [MODERN PERIOD], engaging with [MAJOR THEOLOGICAL FIGURE 1] and [MAJOR THEOLOGICAL FIGURE 2]. How do contemporary challenges like [MODERN ISSUE] require further doctrinal development or restatement?

4. **Confessional Theology Engagement**: Analyze how [CONFESSIONAL DOCUMENT] addresses [THEOLOGICAL ISSUE], comparing its approach with [CONTEMPORARY THEOLOGICAL MOVEMENT]. What aspects of this confessional position remain binding for [DENOMINATIONAL CONTEXT], and where might theological development be appropriate?

5. **Trinitarian Theology Application**: Examine how Trinitarian theology illuminates [SPECIFIC THEOLOGICAL/PRACTICAL ISSUE], engaging with both [CLASSICAL TRINITARIAN THEOLOGIAN] and [CONTEMPORARY TRINITARIAN THEOLOGIAN]. How does Trinitarian doctrine provide resources for [PASTORAL/ETHICAL CHALLENGE]?

6. **Christological Analysis**: Develop a Christological response to [CONTEMPORARY CHALLENGE], drawing on [CLASSICAL CHRISTOLOGICAL COUNCIL/THEOLOGIAN] and engaging with [MODERN CHRISTOLOGICAL DEBATE]. How does the person and work of Christ address [SPECIFIC

PASTORAL/THEOLOGICAL CONCERN]?

7. **Pneumatological Integration**: Examine the role of the Holy Spirit in [THEOLOGICAL/PRACTICAL AREA], comparing [TRADITIONAL PNEUMATOLOGY] with [CONTEMPORARY PNEUMATOLOGICAL EMPHASIS]. How might renewed attention to pneumatology inform [MINISTERIAL PRACTICE/THEOLOGICAL UNDERSTANDING]?

8. **Anthropological Theology**: Construct a theological anthropology that addresses [CONTEMPORARY CHALLENGE TO HUMAN DIGNITY], engaging with [THEOLOGICAL TRADITION] and [SECULAR ANTHROPOLOGICAL PERSPECTIVE]. What does it mean to be created in the image of God in light of [SPECIFIC MODERN DEVELOPMENT]?

9. **Ecclesiological Development**: Analyze how [ECCLESIOLOGICAL MODEL/THEOLOGIAN] might inform the church's response to [CONTEMPORARY ECCLESIAL CHALLENGE]. What theological principles should guide

[CHURCH STRUCTURE/PRACTICE/MISSION] in the context of [CULTURAL/SOCIAL SITUATION]?

10. **Eschatological Integration**: Examine how eschatological hope shapes Christian engagement with [TEMPORAL CONCERN], comparing [ESCHATOLOGICAL POSITION 1] with [ESCHATOLOGICAL POSITION 2]. How does [ESCHATOLOGICAL DOCTRINE] inform [ETHICAL/PASTORAL RESPONSE] to [SPECIFIC CHALLENGE]?

11. **Sacramental Theology**: Develop a sacramental theology of [SPECIFIC SACRAMENT] that addresses [CONTEMPORARY PASTORAL/LITURGICAL ISSUE], engaging with [TRADITIONAL SACRAMENTAL THEOLOGY] and [MODERN SACRAMENTAL DISCUSSION]. How might sacramental practice be renewed to serve [PASTORAL GOAL]?

12. **Theological Aesthetics**: Examine how [THEOLOGICAL TRADITION] understands the relationship between

[DIVINE BEAUTY/TRUTH/GOODNESS] and [ARTISTIC/CULTURAL EXPRESSION]. How might theological aesthetics inform [WORSHIP/CULTURAL ENGAGEMENT/PASTORAL CARE]?

13. **Contextual Theology Construction**: Develop a contextual theological response to [LOCAL/CULTURAL CHALLENGE] that draws on [GLOBAL THEOLOGICAL RESOURCE] while remaining faithful to [CONFESSIONAL/SCRIPTURAL FOUNDATION]. How does [CONTEXTUAL FACTOR] require theological adaptation without compromising [ESSENTIAL DOCTRINE]?

14. **Interfaith Theological Engagement**: Construct a theological framework for Christian engagement with [SPECIFIC RELIGION] regarding [SHARED CONCERN], drawing on [THEOLOGICAL RESOURCE 1] and [THEOLOGICAL RESOURCE 2]. How does [CHRISTIAN DOCTRINE] inform both respectful dialogue and faithful witness?

15. **Public Theology Development**: Develop a public theological position on [POLICY/SOCIAL ISSUE] that engages [SECULAR DISCOURSE] while

maintaining [THEOLOGICAL INTEGRITY]. How can [CHRISTIAN DOCTRINE/ETHICS] contribute to [PUBLIC DEBATE] without either triumphalism or accommodation?

Biblical Exegesis and Hermeneutics (15 prompts)

16. **Advanced Exegetical Analysis**: Conduct a thorough exegetical analysis of [BIBLICAL PASSAGE] using [EXEGETICAL METHOD 1] and [EXEGETICAL METHOD 2], with particular attention to [THEOLOGICAL/PASTORAL QUESTION]. How do different methodological approaches illuminate various dimensions of the text's meaning for [CONTEMPORARY APPLICATION]?

17. **Canonical Hermeneutics**: Examine how [SPECIFIC BIBLICAL TEXT] functions within the broader canonical context of [TESTAMENT/CORPUS], particularly its relationship to [RELATED BIBLICAL THEME/TEXT]. How does canonical reading affect interpretation of [DISPUTED/COMPLEX PASSAGE] for [DOCTRINAL/PASTORAL PURPOSE]?

18. **Theological Interpretation**: Apply theological interpretation methods to [BIBLICAL BOOK/PASSAGE], engaging with [PATRISTIC INTERPRETER] and [CONTEMPORARY THEOLOGICAL INTERPRETER]. How does reading this text within the rule of faith illuminate [DOCTRINAL/SPIRITUAL INSIGHT] for [CONTEMPORARY CHURCH CONTEXT]?

19. **Cultural-Critical Analysis**: Examine [BIBLICAL TEXT] through [CULTURAL-CRITICAL LENS], analyzing how [ANCIENT CULTURAL CONTEXT] and [CONTEMPORARY CULTURAL SITUATION] affect interpretation. What hermeneutical principles should guide application of this text to [SPECIFIC PASTORAL/ETHICAL SITUATION]?

20. **Intertextual Exegesis**: Analyze the intertextual relationships between [BIBLICAL TEXT 1] and [BIBLICAL TEXT 2], examining how [LATER TEXT] reinterprets or develops [EARLIER TEXT]. What theological development does this intertextual relationship reveal about [BIBLICAL THEME/DOCTRINE]?

21. **Narrative Theology Approach**: Apply narrative theological methods to [BIBLICAL NARRATIVE], examining how [PLOT/CHARACTER/SETTING] elements contribute to theological meaning. How does this narrative shape Christian identity and practice regarding [SPECIFIC ASPECT OF DISCIPLESHIP]?

22. **Wisdom Literature Integration**: Examine [WISDOM TEXT] within the broader context of biblical wisdom traditions, analyzing its approach to [LIFE QUESTION/THEOLOGICAL PROBLEM]. How does biblical wisdom inform [PASTORAL CARE/CHRISTIAN LIVING] in the context of [CONTEMPORARY CHALLENGE]?

23. **Apocalyptic Hermeneutics**: Develop hermeneutical principles for interpreting [APOCALYPTIC TEXT] that balance historical context with theological significance. How should contemporary Christians understand this text's message about [DIVINE ACTION/ESCHATOLOGICAL HOPE] in light of [CURRENT SITUATION]?

24. **Prophetic Literature Analysis**: Examine [PROPHETIC TEXT] as both historical

document and theological resource, analyzing its critique of [SOCIAL/RELIGIOUS ISSUE]. How does this prophetic message inform contemporary Christian engagement with [ANALOGOUS MODERN ISSUE]?

25. **Pauline Theology Integration**: Analyze [PAULINE PASSAGE] within the broader context of Pauline theology, examining its contribution to understanding [PAULINE THEME]. How does this text inform contemporary Christian understanding of [DOCTRINAL/PRACTICAL ISSUE]?

26. **Gospel Studies Methodology**: Compare the treatment of [SHARED EPISODE/THEME] across [SYNOPTIC GOSPELS/JOHN], using redaction criticism to identify each evangelist's theological emphasis. What pastoral and theological insights emerge from these different presentations?

27. **Hebrew Bible Theology**: Examine [OLD TESTAMENT TEXT] within the context of Hebrew Bible theology, analyzing its understanding of [THEOLOGICAL THEME]. How does this text contribute to Christian theological reflection on

[RELATED NEW TESTAMENT/DOCTRINAL TOPIC]?

28. **Liturgical Hermeneutics**: Analyze how [BIBLICAL TEXT] functions in [LITURGICAL CONTEXT], examining the relationship between exegetical meaning and liturgical use. How does liturgical context illuminate aspects of the text relevant for [WORSHIP/SPIRITUAL FORMATION]?

29. **Contextual Biblical Interpretation**: Examine how [CONTEXTUAL HERMENEUTICAL APPROACH] interprets [BIBLICAL TEXT], comparing this reading with [TRADITIONAL INTERPRETATION]. What legitimate insights does this approach offer while maintaining [DOCTRINAL/CONFESSIONAL BOUNDARIES]?

30. **Homiletical Exegesis**: Develop a homiletical approach to [BIBLICAL TEXT] that maintains exegetical integrity while addressing [CONTEMPORARY PASTORAL CONCERN]. How can faithful exposition of this text speak to [SPECIFIC CONGREGATIONAL/CULTURAL SITUATION]?

Historical Theology and Church History (15 prompts)

31. **Patristic Theology Engagement**: Analyze [PATRISTIC THEOLOGIAN]'s treatment of [THEOLOGICAL ISSUE], examining both historical context and contemporary relevance. How might [CHURCH FATHER/MOTHER]'s insights inform current theological discussion of [RELATED MODERN ISSUE]?

32. **Medieval Synthesis Study**: Examine [MEDIEVAL THEOLOGIAN]'s integration of [PHILOSOPHICAL SYSTEM] with [THEOLOGICAL TRADITION], focusing on their approach to [SPECIFIC ISSUE]. What methodological insights from this synthesis remain valuable for contemporary theological work?

33. **Reformation Theology Analysis**: Analyze [REFORMATION FIGURE]'s position on [THEOLOGICAL/ECCLESIAL ISSUE], comparing their approach with [CONTEMPORARY POSITION]. How do Reformation insights inform current debates about [RELATED MODERN QUESTION]?

34. **Modern Theology Engagement**: Examine [MODERN THEOLOGIAN]'s response to [ENLIGHTENMENT/MODERN CHALLENGE], analyzing both their theological method and conclusions. What aspects of their approach remain viable for addressing [CONTEMPORARY THEOLOGICAL CHALLENGE]?

35. **Conciliar Theology Study**: Analyze the theological development evident in [ECUMENICAL COUNCIL], examining the issues that prompted conciliar action and the theological solutions proposed. How do conciliar principles inform contemporary approaches to [RELATED THEOLOGICAL/ECCLESIAL ISSUE]?

36. **Denominational History Analysis**: Trace the development of [DENOMINATIONAL TRADITION] from [FOUNDING PERIOD] through [CONTEMPORARY PERIOD], focusing on [DISTINCTIVE THEOLOGICAL/PRACTICAL EMPHASIS]. How has this tradition's approach to [SPECIFIC ISSUE] evolved, and what factors have driven change?

37. **Revival and Renewal Movements**: Examine [REVIVAL/RENEWAL MOVEMENT] within [HISTORICAL

CONTEXT], analyzing its theological emphases and lasting impact. What factors contributed to this movement's emergence, and how did it reshape understanding of [SPECIFIC ASPECT OF FAITH/PRACTICE]?

38. **Theological Controversy Study**: Analyze the [THEOLOGICAL CONTROVERSY] between [OPPOSING POSITIONS], examining the underlying theological and cultural factors. What hermeneutical and theological principles emerged from this controversy that inform contemporary theological method?

39. **Contextual Church History**: Examine the development of Christianity in [GEOGRAPHICAL/CULTURAL CONTEXT], analyzing how [LOCAL FACTORS] shaped theological and ecclesial development. What insights from this contextual development inform contemporary [MISSION/CONTEXTUALIZATION] efforts?

40. **Women in Church History**: Analyze the contributions of [FEMALE CHURCH FIGURE] to [THEOLOGICAL/SPIRITUAL/INSTITUTI

ONAL DEVELOPMENT], examining both her achievements and the obstacles she faced. How does her example inform contemporary discussions about [WOMEN'S ROLES/LEADERSHIP/THEOLOGY]?

41. **Liturgical Development Study**: Trace the historical development of [LITURGICAL PRACTICE/SEASON], examining how theological understanding and pastoral needs shaped liturgical evolution. What principles should guide contemporary liturgical [RENEWAL/ADAPTATION]?

42. **Theological Education History**: Examine the development of [THEOLOGICAL EDUCATION MODEL/INSTITUTION], analyzing how educational philosophy and social context shaped theological formation. What insights inform contemporary discussions about [MINISTERIAL PREPARATION/THEOLOGICAL CURRICULUM]?

43. **Missions History Analysis**: Analyze [MISSIONARY MOVEMENT/FIGURE] within [HISTORICAL/CULTURAL CONTEXT], examining both achievements and failures in [CROSS-CULTURAL MISSION]. What lessons inform

contemporary approaches to [GLOBAL MISSION/CULTURAL ENGAGEMENT]?

44. **Social Justice in Church History**: Examine how [CHURCH FIGURE/MOVEMENT] addressed [SOCIAL JUSTICE ISSUE] within [HISTORICAL CONTEXT], analyzing theological motivations and practical strategies. How does this example inform contemporary Christian engagement with [RELATED MODERN ISSUE]?

45. **Theological Method Development**: Trace the evolution of [THEOLOGICAL METHOD/APPROACH] from [HISTORICAL PERIOD] to [CONTEMPORARY PERIOD], examining key figures and developments. How has this methodological tradition adapted to address [MODERN THEOLOGICAL CHALLENGES]?

Pastoral Theology and Practical Ministry (15 prompts)

46. **Pastoral Care Integration**: Develop a theological framework for pastoral care that addresses [SPECIFIC CARE SITUATION], integrating insights from [THEOLOGICAL

TRADITION], [PSYCHOLOGICAL PERSPECTIVE], and [PASTORAL THEOLOGY RESOURCE]. How do theological convictions shape [CARE APPROACH/INTERVENTION STRATEGY]?

47. **Preaching Theology Development**: Construct a theology of preaching that addresses [HOMILETICAL CHALLENGE], drawing on [BIBLICAL THEOLOGY], [HISTORICAL PREACHING TRADITION], and [CONTEMPORARY HOMILETICAL THEORY]. How should theological convictions about [DOCTRINE/THEME] inform [SERMON PREPARATION/DELIVERY]?

48. **Liturgical Theology Application**: Develop liturgical practices for [WORSHIP CONTEXT/OCCASION] that reflect sound theological principles while addressing [PASTORAL/CULTURAL CONCERN]. How do theological convictions about [WORSHIP/SACRAMENTS] inform [SPECIFIC LITURGICAL DECISION]?

49. **Christian Education Framework**: Design a Christian education approach for [AGE GROUP/CONTEXT] that integrates

[EDUCATIONAL THEORY] with [THEOLOGICAL FOUNDATION]. How should theological understanding of [HUMAN DEVELOPMENT/FAITH FORMATION] shape [PEDAGOGICAL METHOD/CURRICULUM]?

50. **Counseling Integration Model**: Develop an integration model for Christian counseling that addresses [PSYCHOLOGICAL/SPIRITUAL ISSUE], combining insights from [THERAPEUTIC APPROACH] and [THEOLOGICAL RESOURCE]. What theological boundaries and opportunities exist in [SPECIFIC COUNSELING SITUATION]?

51. **Leadership Theology**: Construct a theological understanding of Christian leadership that addresses [LEADERSHIP CHALLENGE], drawing on [BIBLICAL LEADERSHIP MODEL] and [CONTEMPORARY LEADERSHIP THEORY]. How do theological convictions about [AUTHORITY/SERVICE/COMMUNITY] inform [LEADERSHIP PRACTICE]?

52. **Spiritual Formation Design**: Develop a spiritual formation program for [TARGET GROUP] that addresses [SPIRITUAL

GROWTH NEED], integrating [SPIRITUAL TRADITION] with [CONTEMPORARY FORMATION APPROACH]. How does theological understanding of [SPIRITUAL DEVELOPMENT] inform [FORMATIONAL PRACTICE]?

53. **Crisis Ministry Framework**: Create a pastoral response framework for [CRISIS SITUATION] that integrates [THEOLOGICAL UNDERSTANDING] with [CRISIS INTERVENTION PRINCIPLES]. How do theological convictions about [SUFFERING/HOPE/DIVINE PRESENCE] inform [PASTORAL RESPONSE/CARE STRATEGY]?

54. **Evangelism Theology**: Develop a theological approach to evangelism that addresses [CULTURAL/CONTEXTUAL CHALLENGE], balancing [GOSPEL FIDELITY] with [CULTURAL SENSITIVITY]. How does theological understanding of [GOSPEL/CULTURE/CONVERSION] inform [EVANGELISTIC METHOD/APPROACH]?

55. **Church Discipline Framework**: Construct a theological and pastoral approach to church discipline that addresses [BEHAVIORAL/DOCTRINAL ISSUE], balancing [CHURCH PURITY] with [PASTORAL CARE]. How do theological convictions about [SIN/FORGIVENESS/RESTORATION] inform [DISCIPLINARY PROCESS]?

56. **Worship Planning Integration**: Develop worship planning principles that integrate [THEOLOGICAL UNDERSTANDING] with [CULTURAL/PASTORAL CONTEXT], addressing [WORSHIP CHALLENGE/OPPORTUNITY]. How should theological convictions about [WORSHIP/LITURGY] inform [WORSHIP DESIGN/LEADERSHIP]?

57. **Youth Ministry Theology**: Create a theological foundation for youth ministry that addresses [DEVELOPMENTAL/CULTURAL CHALLENGE], integrating [THEOLOGICAL ANTHROPOLOGY] with [YOUTH DEVELOPMENT UNDERSTANDING]. How do theological convictions about [ADOLESCENCE/FAITH

DEVELOPMENT] inform [YOUTH MINISTRY PRACTICE]?

58. **Marriage and Family Ministry**: Develop a pastoral approach to [MARRIAGE/FAMILY ISSUE] that integrates [THEOLOGICAL UNDERSTANDING] with [FAMILY SYSTEMS THEORY]. How do theological convictions about [MARRIAGE/FAMILY/RELATIONSHIPS] inform [PASTORAL INTERVENTION/SUPPORT]?

59. **Social Justice Ministry**: Design a congregational approach to [SOCIAL JUSTICE ISSUE] that integrates [THEOLOGICAL FOUNDATION] with [COMMUNITY ORGANIZING PRINCIPLES]. How do theological convictions about [JUSTICE/COMPASSION/TRANSFORMATION] inform [SOCIAL ACTION/ADVOCACY]?

60. **Interfaith Pastoral Care**: Develop pastoral care approaches for [MULTIFAITH/MULTICULTURAL CONTEXT], balancing [CHRISTIAN IDENTITY] with [RELIGIOUS SENSITIVITY]. How do theological

convictions about [HUMAN DIGNITY/RELIGIOUS DIFFERENCE] inform [CHAPLAINCY/COMMUNITY MINISTRY]?

Ethics and Moral Theology (15 prompts)

61. **Bioethics Integration**: Develop a Christian ethical framework for [BIOETHICAL DILEMMA] that integrates [THEOLOGICAL ANTHROPOLOGY], [MEDICAL KNOWLEDGE], and [ETHICAL THEORY]. How do theological convictions about [HUMAN LIFE/DIGNITY/STEWARDSHIP] inform decision-making regarding [SPECIFIC MEDICAL SITUATION]?

62. **Environmental Ethics Construction**: Create a theological approach to [ENVIRONMENTAL ISSUE] that draws on [CREATION THEOLOGY], [ECOLOGICAL SCIENCE], and [ENVIRONMENTAL ETHICS THEORY]. How do theological convictions about [CREATION CARE/STEWARDSHIP] inform [ENVIRONMENTAL POLICY/PERSONAL PRACTICE]?

63. **Economic Ethics Framework**: Develop a Christian economic ethic that addresses [ECONOMIC SYSTEM/ISSUE], integrating [THEOLOGICAL UNDERSTANDING] with [ECONOMIC

THEORY]. How do theological convictions about [JUSTICE/STEWARDSHIP/COMMUNITY] inform [ECONOMIC POLICY/BUSINESS PRACTICE]?

64. **Political Ethics Analysis**: Construct a Christian approach to [POLITICAL ISSUE/SYSTEM] that balances [THEOLOGICAL CONVICTION] with [DEMOCRATIC PRINCIPLE]. How do theological understandings of [AUTHORITY/JUSTICE/COMMON GOOD] inform [POLITICAL ENGAGEMENT/POLICY POSITION]?

65. **Sexual Ethics Integration**: Develop a theological sexual ethic that addresses [SEXUALITY ISSUE], integrating [BIBLICAL TEACHING], [THEOLOGICAL TRADITION], and [CONTEMPORARY UNDERSTANDING]. How do theological convictions about [EMBODIMENT/RELATIONSHIPS/COVENANT] inform [PASTORAL/ETHICAL GUIDANCE]?
66. **Technology Ethics Framework**: Create a Christian ethical approach to [TECHNOLOGICAL DEVELOPMENT], considering implications for [HUMAN FLOURISHING/COMMUNITY/DIGNITY]

. How do theological convictions about [HUMAN NATURE/TECHNOLOGY/STEWARDSHIP] inform [TECHNOLOGY USE/POLICY]?

67. **Business Ethics Application**: Develop Christian principles for [BUSINESS PRACTICE/DECISION], integrating [THEOLOGICAL UNDERSTANDING] with [BUSINESS ETHICS THEORY]. How do theological convictions about [INTEGRITY/JUSTICE/SERVICE] inform [CORPORATE RESPONSIBILITY/PROFESSIONAL PRACTICE]?

68. **Medical Ethics Integration**: Construct a theological approach to [MEDICAL ETHICS ISSUE] that addresses [HEALTHCARE CHALLENGE], balancing [THEOLOGICAL PRINCIPLE] with [MEDICAL IMPERATIVE]. How do theological convictions about [HEALING/SUFFERING/HOPE] inform [HEALTHCARE PRACTICE/POLICY]?

69. **Social Ethics Development**: Create a Christian social ethic that addresses [SOCIAL ISSUE], drawing on [THEOLOGICAL RESOURCE] and

[SOCIAL ANALYSIS]. How do theological convictions about [HUMAN RELATIONSHIP/COMMUNITY/JUSTICE] inform [SOCIAL POLICY/COMMUNITY ENGAGEMENT]?

70. **International Ethics Framework**: Develop a Christian approach to [GLOBAL ISSUE], integrating [THEOLOGICAL UNDERSTANDING] with [INTERNATIONAL RELATIONS THEORY]. How do theological convictions about [UNIVERSAL DIGNITY/GLOBAL SOLIDARITY] inform [FOREIGN POLICY/INTERNATIONAL ENGAGEMENT]?

71. **Professional Ethics Integration**: Construct ethical guidelines for [PROFESSIONAL CONTEXT] that integrate [CHRISTIAN CONVICTION] with [PROFESSIONAL STANDARD]. How do theological understandings of [VOCATION/INTEGRITY/SERVICE] inform [PROFESSIONAL PRACTICE/DECISION-MAKING]?

72. **Criminal Justice Ethics**: Develop a Christian approach to [CRIMINAL JUSTICE ISSUE] that balances [JUSTICE] with [MERCY/RESTORATION]. How do

theological convictions about [SIN/FORGIVENESS/TRANSFORMATION] inform [JUSTICE SYSTEM/CORRECTIONAL PRACTICE]?

73. **Research Ethics Framework**: Create ethical guidelines for [RESEARCH CONTEXT] that integrate [THEOLOGICAL PRINCIPLE] with [RESEARCH METHODOLOGY]. How do theological convictions about [TRUTH/HUMAN DIGNITY/KNOWLEDGE] inform [RESEARCH PRACTICE/ACADEMIC INTEGRITY]?

74. **Media Ethics Application**: Develop a Christian approach to [MEDIA/COMMUNICATION ISSUE] that addresses [INFORMATION/INFLUENCE CHALLENGE]. How do theological convictions about [TRUTH/COMMUNICATION/COMMUNITY] inform [MEDIA PRACTICE/CONSUMPTION]?

75. **End-of-Life Ethics**: Construct a theological framework for [END-OF-LIFE ISSUE] that integrates [THEOLOGICAL UNDERSTANDING] with [MEDICAL/LEGAL CONSIDERATION].

How do theological convictions about [LIFE/DEATH/HOPE] inform [PASTORAL CARE/ETHICAL DECISION-MAKING]?

Advanced Usage Guidelines for M.Div. and Theological Studies

Substitution Categories for Graduate-Level Work:

[THEOLOGICAL TRADITION]: Reformed, Lutheran, Anglican, Catholic, Orthodox, Wesleyan, Pentecostal, Baptist, etc.

[CLASSICAL THEOLOGIAN]: Augustine, Aquinas, Luther, Calvin, Wesley, Edwards, Barth, Rahner, etc.

[CONTEMPORARY THEOLOGIAN]: Moltmann, Pannenberg, Gunton, LaCugna, Johnson, Volf, etc.

[BIBLICAL BOOK/PASSAGE]: Specific texts relevant to your exegetical focus
[DOCTRINE]: Trinity, Christology, Soteriology, Ecclesiology, Eschatology, etc.

[PASTORAL CONTEXT]: Urban ministry, rural church, hospital chaplaincy, campus ministry, etc.

[ETHICAL ISSUE]: Climate change, economic inequality, reproductive ethics, warfare, etc.

[HISTORICAL PERIOD]: Patristic, Medieval, Reformation, Modern, Contemporary
[CULTURAL CONTEXT]: African, Asian, Latin American, Indigenous, Western, etc.

Academic Integration Strategies:

1. **Primary Source Engagement**: Always engage primary theological sources rather than secondary summaries

2. **Methodological Rigor**: Apply appropriate theological, exegetical, historical, and pastoral methodologies

3. **Interdisciplinary Integration**: Connect theological reflection with relevant insights from other academic disciplines

4. **Confessional Awareness**: Maintain awareness of your own theological tradition while engaging others fairly

5. **Pastoral Application**: Connect academic theology with practical ministry concerns and applications

6. **Contemporary Relevance**: Address how historical theology and biblical exegesis inform current church and world situations

Graduate-Level Expectations:

- **Thesis-Quality Research**: Each prompt can support 50-100+ page research projects
- **Comprehensive Literature Review**: Engage both classical and contemporary sources extensively
- **Original Theological Contribution**: Move beyond summary to constructive theological work
- **Methodological Sophistication**: Apply advanced theological, exegetical, and historical methods
- **Pastoral Integration**: Connect academic work with ministerial preparation and practice
- **Denominational/Confessional Context**: Work within and from your particular theological tradition

Specialized Applications:

- **Master's Thesis Development**: Use as frameworks for major research projects

- **Doctoral Comprehensive Exams**: Practice systematic theological analysis across disciplines

- **Professional Ministry Preparation**: Connect academic theology with practical ministry challenges

- **Academic Conference Presentations**: Develop focused research for scholarly venues

- **Denominational Position Papers**: Contribute to church policy and theological discussion

- **Teaching Preparation**: Develop expertise for theological education and church education ministries

These prompts are designed to bridge the gap between academic theological study and practical ministerial preparation, encouraging both scholarly rigor and pastoral sensitivity appropriate for graduate theological education.

75 Customizable AI Prompts for Doctoral Theological Research (D.D., Ph.D., D.Min.)

75 Customizable AI Prompts for Doctoral Theological Research (D.D., Ph.D., D.Min.)

75 customizable AI prompts specifically designed for doctoral-level theological research (D.D., Ph.D., D.Min.). These prompts reflect the highest academic standards and original research expectations appropriate for terminal degrees in theology and ministry.

Key Features of These Doctoral-Level Prompts:

1. **Original Research Emphasis**: Each prompt is designed to support doctoral dissertations, scholarly publications, and significant contributions to theological knowledge or ministry practice.

2. **Methodological Sophistication**: Prompts require advanced application of theological, historical, exegetical, and interdisciplinary research methodologies.

3. **Three Doctoral Contexts**:
   - **Ph.D.**: Academic research contributing to theological scholarship

  - **D.D.**: Constructive theology for church and academy
  - **D.Min.**: Practical theological research improving ministry effectiveness

4. **Interdisciplinary Integration**: Advanced engagement with non-theological disciplines while maintaining theological integrity.

5. **Global Scholarly Engagement**: Expectation of engagement with international theological discourse and diverse cultural contexts.

The Five Categories Include:

- **Systematic and Constructive Theology Research** (15 prompts): For original theological construction and doctrinal innovation

- **Advanced Biblical and Textual Studies** (15 prompts): For cutting-edge exegetical and hermeneutical research

- **Historical Theology and Intellectual History** (15 prompts): For comprehensive historical analysis and recovery of theological traditions

- **Practical Theology and Ministry Studies** (15 prompts): For research improving ministry practice and church effectiveness

- **Ethics, Society, and Applied Theology** (15 prompts): For addressing contemporary ethical challenges through theological analysis

Doctoral-Level Distinctions:

These prompts differ from master's level by requiring:

- **Original contribution** to theological knowledge or ministry practice
- **Comprehensive literature review** of international scholarship
- **Methodological innovation** advancing research approaches
- **Interdisciplinary sophistication** integrating multiple academic fields
- **Publication-quality research** suitable for peer-reviewed venues
- **Global theological awareness** engaging diverse cultural contexts

Research Applications:

- **Doctoral Dissertations**: Complete 200-400+ page research projects

- **Scholarly Publications**: Peer-reviewed journal articles and academic books
- **Academic Conferences**: Research presentations for scholarly societies
- **Grant Applications**: Research proposals for academic or ministry funding
- **Professional Consultation**: Expert analysis for denominational or organizational challenges
- **Postdoctoral Research**: Advanced research for academic career development

Quality Standards:

Each prompt is designed to support research that meets the standards for:

- Academic publication in peer-reviewed journals
- Presentation at major theological conferences
- Denominational theological consultation
- International scholarly engagement
- Professional ministry advancement
- Academic career preparation

These prompts provide the framework for the most advanced theological research while maintaining flexibility for diverse research interests, denominational contexts, and career goals in theological education and ministry leadership.

75 Customizable AI Prompts for Doctoral Theological Research (D.D., Ph.D., D.Min.)

Instructions: Replace bracketed placeholders with your specific research focus. These prompts are designed for original doctoral-level research requiring comprehensive literature review, methodological sophistication, and significant contribution to theological scholarship or ministerial practice.

Systematic and Constructive Theology Research (15 prompts)

1. **Constructive Theological Proposal**: Develop an original systematic theological construction of [DOCTRINE/CONCEPT] that addresses [CONTEMPORARY THEOLOGICAL/CULTURAL CHALLENGE] through critical engagement with [CLASSICAL THEOLOGICAL TRADITION], [CONTEMPORARY THEOLOGICAL MOVEMENT], and [INTERDISCIPLINARY RESOURCE]. What new theological insights emerge from

this synthesis, and how do they advance theological understanding of [FUNDAMENTAL THEOLOGICAL QUESTION]?

2. **Theological Method Innovation**: Propose and defend an innovative theological methodology that integrates [METHODOLOGICAL APPROACH 1], [METHODOLOGICAL APPROACH 2], and [INTERDISCIPLINARY METHOD] for addressing [COMPLEX THEOLOGICAL PROBLEM]. How does this methodological innovation overcome limitations in existing approaches to [THEOLOGICAL INQUIRY AREA], and what criteria validate its theological adequacy?

3. **Doctrinal Reconstruction**: Undertake a critical reconstruction of [TRADITIONAL DOCTRINE] in dialogue with [POSTMODERN/CONTEMPORARY CHALLENGE], engaging extensively with [HISTORICAL DEVELOPMENT] and [CURRENT THEOLOGICAL DEBATE]. What aspects of traditional formulation require reformulation, and how does your reconstruction maintain theological continuity while addressing [CONTEMPORARY INADEQUACY]?

4. **Theological Correlation Analysis**: Develop a correlational theology that brings [THEOLOGICAL TRADITION] into sustained dialogue with [NON-THEOLOGICAL DISCIPLINE/WORLDVIEW], examining both convergences and irreducible differences. How does this correlation illuminate [THEOLOGICAL TOPIC] while maintaining theological integrity and advancing interdisciplinary understanding?

5. **Contextual Theological Innovation**: Create an original contextual theological framework for [SPECIFIC CULTURAL/GEOGRAPHICAL CONTEXT] that critically appropriates [GLOBAL THEOLOGICAL RESOURCE] while addressing [LOCAL THEOLOGICAL/PRACTICAL CHALLENGE]. What methodological principles govern authentic contextualization, and how does your framework contribute to global theological discourse?

6. **Theological Anthropology Development**: Construct a comprehensive theological anthropology that engages [SCIENTIFIC/PHILOSOPHICAL ANTHROPOLOGY], [CLASSICAL

THEOLOGICAL ANTHROPOLOGY], and [CONTEMPORARY CULTURAL ANALYSIS]. How does your anthropological construction address [MODERN CHALLENGE TO HUMAN DIGNITY/MEANING], and what implications follow for [SOTERIOLOGY/ETHICS/ECCLESIOLOGY]?

7. **Trinitarian Theology Advancement**: Develop an original contribution to Trinitarian theology that addresses [CONTEMPORARY THEOLOGICAL/PHILOSOPHICAL CHALLENGE] through engagement with [PATRISTIC SOURCES], [CONTEMPORARY TRINITARIAN THEOLOGIANS], and [PHILOSOPHICAL/SCIENTIFIC RESOURCE]. How does your Trinitarian construction illuminate [THEOLOGICAL/PRACTICAL PROBLEM] and advance Trinitarian discourse?

8. **Eschatological Theology Construction**: Propose a comprehensive eschatological theology that addresses [CONTEMPORARY CHALLENGE] through critical engagement with

[BIBLICAL ESCHATOLOGY], [HISTORICAL ESCHATOLOGICAL POSITIONS], and [CONTEMPORARY ESCHATOLOGICAL DEBATE]. How does your eschatological framework inform [CHRISTIAN ETHICS/MISSION/HOPE] in the context of [MODERN SITUATION]?

9. **Pneumatological Innovation**: Develop an original pneumatological contribution that addresses [THEOLOGICAL/PRACTICAL GAP] through engagement with [PNEUMATOLOGICAL TRADITION], [CONTEMPORARY PNEUMATOLOGICAL THEOLOGY], and [EXPERIENTIAL/CHARISMATIC RESOURCE]. How does renewed pneumatological understanding transform approaches to [THEOLOGICAL AREA/MINISTERIAL PRACTICE]?

10. **Ecological Theology Construction**: Create a comprehensive theological framework for [ENVIRONMENTAL/ECOLOGICAL ISSUE] that integrates [CREATION THEOLOGY], [ECOLOGICAL SCIENCE], and [ENVIRONMENTAL PHILOSOPHY]. How does your ecological theology address [ENVIRONMENTAL CRISIS] while maintaining theological distinctiveness and practical effectiveness?

11. **Theological Epistemology Development**: Propose an original theological epistemology that addresses [EPISTEMOLOGICAL CHALLENGE] through engagement with [CLASSICAL THEOLOGICAL EPISTEMOLOGY], [CONTEMPORARY PHILOSOPHY OF KNOWLEDGE], and [RELIGIOUS EXPERIENCE ANALYSIS]. How does your epistemological framework advance theological knowledge claims while addressing [MODERN SKEPTICISM/PLURALISM]?

12. **Public Theology Innovation**: Develop a sophisticated public theological framework for [PUBLIC POLICY/SOCIAL ISSUE] that maintains theological integrity while engaging [SECULAR DISCOURSE] and [DEMOCRATIC PLURALISM]. How does your public theological approach contribute to [PUBLIC DEBATE] while avoiding both sectarian withdrawal and theological accommodation?

13. **Theological Aesthetics Construction**: Create an original theological aesthetics that addresses [CULTURAL/ARTISTIC PHENOMENON] through engagement with [THEOLOGICAL TRADITION],

[AESTHETIC THEORY], and [CULTURAL ANALYSIS]. How does your theological aesthetics illuminate [DIVINE BEAUTY/CULTURAL ENGAGEMENT] and inform [WORSHIP/CULTURAL CRITICISM]?

14. **Interfaith Theological Framework**: Develop a sophisticated theological framework for [INTERFAITH RELATIONSHIP/DIALOGUE] that maintains [CHRISTIAN THEOLOGICAL DISTINCTIVENESS] while enabling genuine dialogue about [SHARED CONCERN]. How does your framework advance interfaith understanding while preserving theological integrity?

15. **Theological Hermeneutics Innovation**: Propose an original hermeneutical approach that integrates [HERMENEUTICAL TRADITION], [CONTEMPORARY HERMENEUTICAL THEORY], and [CONTEXTUAL INTERPRETIVE CONCERN]. How does your hermeneutical innovation advance theological interpretation while addressing [INTERPRETIVE CHALLENGE/CULTURAL BARRIER]?

Advanced Biblical and Textual Studies (15 prompts)

16. **Comprehensive Exegetical Monograph**: Conduct an exhaustive exegetical analysis of [BIBLICAL BOOK/CORPUS] using advanced critical methodologies, with particular focus on [THEOLOGICAL/LITERARY THEME]. How does your analysis contribute new insights to [BIBLICAL SCHOLARSHIP AREA] and inform [THEOLOGICAL/PASTORAL APPLICATION]?

17. **Hermeneutical Innovation**: Develop an original hermeneutical approach that integrates [TRADITIONAL EXEGETICAL METHOD], [CONTEMPORARY INTERPRETIVE THEORY], and [CONTEXTUAL READING STRATEGY]. How does your hermeneutical innovation address limitations in current approaches to [BIBLICAL INTERPRETATION CHALLENGE]?

18. **Canonical Theology Construction**: Develop a comprehensive canonical theological reading of [BIBLICAL THEME/CORPUS] that traces development

from [EARLY TEXTS] through [CANONICAL COMPLETION]. How does canonical reading reveal theological dimensions unavailable to historical-critical analysis alone, and what implications follow for [SYSTEMATIC THEOLOGY/CHURCH PRACTICE]?

19. **Ancient Near Eastern Integration**: Conduct advanced comparative analysis of [BIBLICAL TEXT/TRADITION] with [ANCIENT NEAR EASTERN SOURCES], employing latest archaeological and philological research. How does this comparative work illuminate [BIBLICAL UNDERSTANDING] while maintaining recognition of [BIBLICAL DISTINCTIVENESS]?

20. **Theological Interpretation Advancement**: Develop a sophisticated theological interpretation of [BIBLICAL CORPUS] that integrates [HISTORICAL-CRITICAL ANALYSIS], [CANONICAL CONTEXT], and [CHRISTIAN DOCTRINAL TRADITION]. How does theological interpretation advance beyond both fundamentalist literalism and critical reductionism?

21. **Reception History Analysis**: Conduct comprehensive reception history analysis of [BIBLICAL TEXT/THEME] from [EARLY PERIOD] through [CONTEMPORARY PERIOD], examining how interpretive traditions have shaped understanding. How has reception history both illuminated and obscured [TEXTUAL MEANING], and what implications follow for current interpretation?

22. **Digital Humanities Application**: Employ advanced digital humanities methodologies to analyze [BIBLICAL/TEXTUAL CORPUS], using [COMPUTATIONAL METHOD] to identify [LINGUISTIC/LITERARY/THEOLOGICAL PATTERNS]. How do digital approaches reveal previously unnoticed dimensions of [BIBLICAL TEXT], and what methodological implications follow?

23. **Manuscript Studies Research**: Conduct original manuscript research on [TEXTUAL TRADITION/VARIANT], employing advanced text-critical methodologies to address [TEXTUAL PROBLEM]. How does your textual analysis contribute to understanding of [TEXTUAL TRANSMISSION/CANONICAL DEVELOPMENT]?

24. **Comparative Canonical Analysis**: Compare canonical treatment of [SHARED TRADITION/TEXT] across [DIFFERENT RELIGIOUS COMMUNITIES], analyzing how canonical decisions affect interpretation. What does this comparison reveal about [CANONICAL FORMATION PROCESS] and [INTERPRETIVE AUTHORITY]?

25. **Linguistic Analysis Innovation**: Employ advanced linguistic analysis of [BIBLICAL LANGUAGE/CORPUS] using [LINGUISTIC METHODOLOGY] to address [INTERPRETIVE/THEOLOGICAL QUESTION]. How does sophisticated linguistic analysis illuminate [SEMANTIC/SYNTACTIC/PRAGMATIC DIMENSIONS] relevant to [THEOLOGICAL UNDERSTANDING]?

26. **Genre Studies Advancement**: Develop comprehensive analysis of [BIBLICAL GENRE] employing advanced literary theory and comparative literature methodology. How does genre analysis illuminate [INTERPRETIVE APPROACH] and inform [THEOLOGICAL/PASTORAL APPLICATION]?

27. **Socio-Rhetorical Analysis**: Conduct advanced socio-rhetorical analysis of [BIBLICAL TEXT/CORPUS] employing [RHETORICAL METHOD] and [SOCIAL-SCIENTIFIC APPROACH]. How does this analysis illuminate [SOCIAL CONTEXT] and [RHETORICAL STRATEGY] relevant to [CONTEMPORARY APPLICATION]?

28. **Wisdom Literature Innovation**: Develop original analysis of [WISDOM CORPUS] that integrates [ANCIENT NEAR EASTERN WISDOM], [BIBLICAL WISDOM TRADITION], and [CONTEMPORARY WISDOM STUDIES]. How does this analysis advance understanding of [WISDOM THEOLOGY] and inform [CONTEMPORARY PRACTICAL THEOLOGY]?

29. **Narrative Theology Development**: Create comprehensive narrative theological analysis of [BIBLICAL NARRATIVE CORPUS] that examines [THEOLOGICAL THEMES] through sophisticated literary analysis. How does narrative approach illuminate [THEOLOGICAL UNDERSTANDING] unavailable through other methodologies?

30. **Intertextual Studies Innovation**: Conduct advanced intertextual analysis examining relationships between [BIBLICAL TEXTS] and [EXTRA-BIBLICAL TRADITIONS], employing [INTERTEXTUAL METHODOLOGY]. How does this analysis illuminate [TEXTUAL DEVELOPMENT] and [THEOLOGICAL MEANING]?

Historical Theology and Intellectual History (15 prompts)

31. **Patristic Studies Innovation**: Conduct original research on [PATRISTIC FIGURE/MOVEMENT] using [PRIMARY SOURCES] and [ADVANCED HISTORICAL METHODOLOGY], addressing [SCHOLARLY DEBATE/GAP]. How does your analysis revise understanding of [PATRISTIC DEVELOPMENT] and inform [CONTEMPORARY THEOLOGICAL ISSUES]?

32. **Medieval Synthesis Analysis**: Develop comprehensive analysis of [MEDIEVAL THEOLOGICAL SYNTHESIS] examining integration of [PHILOSOPHICAL TRADITION] with [THEOLOGICAL TRADITION]. How does this synthesis

inform contemporary attempts at [THEOLOGICAL/PHILOSOPHICAL INTEGRATION]?

33. **Reformation Studies Advancement**: Conduct original research on [REFORMATION FIGURE/MOVEMENT] that addresses [SCHOLARLY CONTROVERSY] through [PRIMARY SOURCE ANALYSIS] and [CONTEXTUAL METHODOLOGY]. How does your research contribute to [REFORMATION SCHOLARSHIP] and inform [CONTEMPORARY ECCLESIASTICAL ISSUES]?

34. **Modern Theology Archaeology**: Undertake archaeological analysis of [MODERN THEOLOGICAL DEVELOPMENT] through comprehensive engagement with [PRIMARY SOURCES] and [INTELLECTUAL CONTEXT]. How does this analysis reveal [HIDDEN INFLUENCES/DEVELOPMENTS] that inform understanding of [CONTEMPORARY THEOLOGICAL SITUATION]?

35. **Denominational History Innovation**: Develop comprehensive denominational history of [ECCLESIASTICAL

TRADITION] that employs [ADVANCED HISTORICAL METHODOLOGY] and addresses [HISTORIOGRAPHICAL DEBATE]. How does your analysis contribute to [DENOMINATIONAL SELF-UNDERSTANDING] and [ECUMENICAL DIALOGUE]?

36. **Theological Biography Construction**: Create definitive theological biography of [THEOLOGICAL FIGURE] that integrates [PERSONAL HISTORY], [INTELLECTUAL DEVELOPMENT], and [CULTURAL CONTEXT]. How does biographical analysis illuminate [THEOLOGICAL CONTRIBUTION] and [HISTORICAL SIGNIFICANCE]?

37. **Conciliar Studies Research**: Conduct original research on [ECUMENICAL/DENOMINATIONAL COUNCIL] employing [CONCILIAR METHODOLOGY] and [PRIMARY SOURCE ANALYSIS]. How does your analysis contribute to understanding of [CONCILIAR AUTHORITY] and [DOCTRINAL DEVELOPMENT]?

38. **Intellectual History Integration**: Develop comprehensive intellectual history of [THEOLOGICAL IDEA/MOVEMENT]

that traces development through [MULTIPLE HISTORICAL PERIODS] and [CULTURAL CONTEXTS]. How does intellectual history analysis illuminate [THEOLOGICAL DEVELOPMENT] and [CONTEMPORARY RELEVANCE]?

39. **Women's History Recovery**: Conduct original research recovering [WOMEN'S THEOLOGICAL CONTRIBUTION] to [HISTORICAL PERIOD/TRADITION] through [FEMINIST HISTORICAL METHODOLOGY]. How does this recovery revise understanding of [THEOLOGICAL DEVELOPMENT] and inform [CONTEMPORARY GENDER DISCUSSIONS]?

40. **Global Christianity Analysis**: Develop comprehensive analysis of [CHRISTIAN MOVEMENT] in [NON-WESTERN CONTEXT] that examines [INDIGENOUS THEOLOGICAL DEVELOPMENT] and [CULTURAL ADAPTATION]. How does this analysis contribute to [GLOBAL THEOLOGICAL UNDERSTANDING] and [POSTCOLONIAL THEOLOGICAL DISCOURSE]?

41. **Liturgical History Research**: Conduct original research on [LITURGICAL

DEVELOPMENT] through [HISTORICAL PERIOD] employing [LITURGICAL HISTORICAL METHODOLOGY]. How does liturgical development reflect [THEOLOGICAL CHANGE] and inform [CONTEMPORARY LITURGICAL RENEWAL]?

42. **Theological Education History**: Develop comprehensive history of [THEOLOGICAL EDUCATION MODEL/INSTITUTION] that analyzes [EDUCATIONAL PHILOSOPHY] and [CULTURAL ADAPTATION]. How does this history inform [CONTEMPORARY THEOLOGICAL EDUCATION CHALLENGES]?

43. **Missions History Innovation**: Conduct original research on [MISSIONARY MOVEMENT] that employs [POSTCOLONIAL METHODOLOGY] and [PRIMARY SOURCE ANALYSIS]. How does this analysis revise understanding of [MISSION HISTORY] and inform [CONTEMPORARY MISSION THEOLOGY]?

44. **Social Christianity Analysis**: Develop comprehensive analysis of [SOCIAL CHRISTIANITY MOVEMENT] that

examines [THEOLOGICAL FOUNDATION] and [SOCIAL ENGAGEMENT]. How does this analysis inform [CONTEMPORARY SOCIAL THEOLOGY] and [CHURCH SOCIAL ENGAGEMENT]?

45. **Theological Methodology History**: Trace historical development of [THEOLOGICAL METHOD] from [ORIGIN PERIOD] through [CONTEMPORARY PERIOD], analyzing [METHODOLOGICAL EVOLUTION]. How does this historical analysis inform [CONTEMPORARY METHODOLOGICAL DEBATES]?

Practical Theology and Ministry Studies (15 prompts)

46. **Pastoral Theology Innovation**: Develop original pastoral theological framework for [MINISTRY CONTEXT/CHALLENGE] that integrates [THEOLOGICAL FOUNDATION], [SOCIAL SCIENTIFIC RESEARCH], and [MINISTERIAL EXPERIENCE]. How does your framework advance [PASTORAL PRACTICE] and contribute to [PRACTICAL THEOLOGICAL DISCOURSE]?

47. **Congregational Studies Research**: Conduct comprehensive congregational analysis employing [ETHNOGRAPHIC METHODOLOGY], [THEOLOGICAL REFLECTION], and [ORGANIZATIONAL ANALYSIS] to address [CONGREGATIONAL CHALLENGE/PHENOMENON]. What insights emerge for [CONGREGATIONAL LEADERSHIP] and [ECCLESIOLOGICAL UNDERSTANDING]?

48. **Homiletical Innovation**: Develop original homiletical theory that addresses [PREACHING CHALLENGE] through integration of [THEOLOGICAL

HERMENEUTICS], [COMMUNICATION THEORY], and [CULTURAL ANALYSIS]. How does your homiletical approach advance [PREACHING EFFECTIVENESS] while maintaining [THEOLOGICAL INTEGRITY]?

49. **Liturgical Studies Development**: Create comprehensive liturgical framework for [WORSHIP CONTEXT] that integrates [LITURGICAL THEOLOGY], [CULTURAL ANTHROPOLOGY], and [AESTHETIC THEORY]. How does your liturgical innovation serve [WORSHIP GOALS] while reflecting [THEOLOGICAL CONVICTIONS]?

50. **Christian Education Innovation**: Develop original Christian education theory that addresses [EDUCATIONAL CHALLENGE] through integration of [THEOLOGICAL ANTHROPOLOGY], [LEARNING THEORY], and [CULTURAL ANALYSIS]. How does your educational approach advance [FAITH FORMATION] in [CONTEMPORARY CONTEXT]?

51. **Pastoral Care Integration**: Create sophisticated pastoral care model that integrates [THEOLOGICAL TRADITION], [PSYCHOLOGICAL THEORY], and

[CULTURAL SENSITIVITY] for addressing [CARE CHALLENGE]. How does your model maintain [THEOLOGICAL DISTINCTIVENESS] while embracing [INTERDISCIPLINARY INSIGHTS]?

52. **Leadership Studies Innovation**: Develop original Christian leadership theory that addresses [LEADERSHIP CHALLENGE] through integration of [THEOLOGICAL UNDERSTANDING], [LEADERSHIP RESEARCH], and [ORGANIZATIONAL DYNAMICS]. How does your leadership model serve [CHRISTIAN MISSION] in [CONTEMPORARY CONTEXT]?

53. **Spiritual Formation Research**: Conduct comprehensive research on [SPIRITUAL FORMATION APPROACH] employing [THEOLOGICAL ANALYSIS], [PSYCHOLOGICAL RESEARCH], and [PHENOMENOLOGICAL METHODOLOGY]. How does your research advance understanding of [SPIRITUAL DEVELOPMENT] and inform [FORMATIONAL PRACTICE]?

54. **Mission Studies Innovation**: Develop original missiological framework that addresses [MISSION CHALLENGE]

through integration of [MISSIONAL THEOLOGY], [CULTURAL ANTHROPOLOGY], and [GLOBAL STUDIES]. How does your framework advance [CROSS-CULTURAL MISSION] while addressing [POSTCOLONIAL CONCERNS]?

55. **Youth Ministry Research**: Conduct comprehensive research on [YOUTH MINISTRY PHENOMENON] employing [THEOLOGICAL REFLECTION], [DEVELOPMENTAL PSYCHOLOGY], and [CULTURAL STUDIES]. How does your research inform [YOUTH MINISTRY PRACTICE] and contribute to [PRACTICAL THEOLOGICAL UNDERSTANDING]?

56. **Digital Ministry Innovation**: Develop original framework for [DIGITAL MINISTRY AREA] that integrates [THEOLOGICAL UNDERSTANDING], [MEDIA THEORY], and [CULTURAL ANALYSIS]. How does your framework address [DIGITAL CHALLENGES] while advancing [MINISTRY EFFECTIVENESS]?

57. **Urban Ministry Studies**: Create comprehensive urban ministry framework

that addresses [URBAN CHALLENGE] through integration of [URBAN THEOLOGY], [SOCIOLOGICAL ANALYSIS], and [COMMUNITY ORGANIZING THEORY]. How does your framework advance [URBAN MISSION] and inform [THEOLOGICAL UNDERSTANDING]?

58. **Interfaith Ministry Research**: Develop original approach to [INTERFAITH MINISTRY CONTEXT] that balances [CHRISTIAN IDENTITY] with [RELIGIOUS PLURALISM]. How does your approach advance [INTERFAITH COOPERATION] while maintaining [THEOLOGICAL INTEGRITY]?

59. **Trauma-Informed Ministry**: Create comprehensive framework for [TRAUMA-INFORMED MINISTRY] that integrates [THEOLOGICAL UNDERSTANDING], [TRAUMA RESEARCH], and [PASTORAL METHODOLOGY]. How does your framework address [TRAUMA RECOVERY] while drawing on [THEOLOGICAL RESOURCES]?

60. **Organizational Ministry Studies**: Develop organizational analysis of [MINISTRY ORGANIZATION] employing

[ORGANIZATIONAL THEORY], [THEOLOGICAL REFLECTION], and [LEADERSHIP STUDIES]. How does your analysis inform [ORGANIZATIONAL EFFECTIVENESS] and [THEOLOGICAL UNDERSTANDING]?

Ethics, Society, and Applied Theology (15 prompts)

61. **Bioethics Innovation**: Develop comprehensive Christian bioethical framework for [BIOETHICAL FRONTIER] that integrates [THEOLOGICAL ANTHROPOLOGY], [MEDICAL SCIENCE], and [ETHICAL THEORY]. How does your framework advance [BIOETHICAL DISCOURSE] while maintaining [THEOLOGICAL DISTINCTIVENESS]?

62. **Environmental Ethics Construction**: Create original environmental theological framework that addresses [ENVIRONMENTAL CRISIS] through integration of [CREATION THEOLOGY], [ECOLOGICAL SCIENCE], and [ENVIRONMENTAL PHILOSOPHY]. How does your framework advance [ENVIRONMENTAL ACTION] and inform [THEOLOGICAL UNDERSTANDING]?

63. **Economic Ethics Development**: Develop sophisticated economic ethical framework that addresses [ECONOMIC CHALLENGE] through integration of

[THEOLOGICAL TRADITION], [ECONOMIC THEORY], and [SOCIAL ANALYSIS]. How does your framework contribute to [ECONOMIC JUSTICE] and [THEOLOGICAL DISCOURSE]?

64. **Political Theology Innovation**: Create original political theological framework for [POLITICAL ISSUE] that integrates [THEOLOGICAL FOUNDATION], [POLITICAL THEORY], and [DEMOCRATIC PRINCIPLES]. How does your framework advance [POLITICAL ENGAGEMENT] while maintaining [THEOLOGICAL INTEGRITY]?

65. **Technology Ethics Research**: Develop comprehensive theological approach to [TECHNOLOGICAL DEVELOPMENT] that addresses [ETHICAL IMPLICATIONS] through integration of [THEOLOGICAL ANTHROPOLOGY] and [TECHNOLOGY STUDIES]. How does your approach guide [TECHNOLOGY USE] while advancing [THEOLOGICAL UNDERSTANDING]?

66. **Global Ethics Construction**: Create sophisticated framework for [GLOBAL ETHICAL ISSUE] that integrates [THEOLOGICAL TRADITION],

[INTERNATIONAL RELATIONS THEORY], and [CULTURAL ANALYSIS]. How does your framework advance [GLOBAL COOPERATION] and inform [THEOLOGICAL DISCOURSE]?

67. **Medical Ethics Integration**: Develop comprehensive medical ethical framework that addresses [HEALTHCARE CHALLENGE] through integration of [THEOLOGICAL UNDERSTANDING], [MEDICAL ETHICS], and [CULTURAL SENSITIVITY]. How does your framework advance [HEALTHCARE PRACTICE] while maintaining [THEOLOGICAL DISTINCTIVENESS]?

68. **Social Justice Innovation**: Create original social justice framework that addresses [SOCIAL INJUSTICE] through integration of [LIBERATION THEOLOGY], [SOCIAL ANALYSIS], and [COMMUNITY ORGANIZING]. How does your framework advance [SOCIAL TRANSFORMATION] and contribute to [THEOLOGICAL DISCOURSE]?

69. **Sexual Ethics Development**: Develop sophisticated sexual ethical framework that addresses [SEXUALITY ISSUE] through integration of [THEOLOGICAL

TRADITION], [SCIENTIFIC UNDERSTANDING], and [CULTURAL ANALYSIS]. How does your framework advance [ETHICAL UNDERSTANDING] while addressing [PASTORAL CONCERNS]?

70. **Criminal Justice Ethics**: Create comprehensive framework for [CRIMINAL JUSTICE REFORM] that integrates [THEOLOGICAL UNDERSTANDING], [CRIMINOLOGICAL RESEARCH], and [RESTORATIVE JUSTICE THEORY]. How does your framework advance [JUSTICE REFORM] and inform [THEOLOGICAL DISCOURSE]?

71. **Business Ethics Innovation**: Develop original business ethical framework that addresses [CORPORATE RESPONSIBILITY ISSUE] through integration of [THEOLOGICAL TRADITION], [BUSINESS ETHICS THEORY], and [STAKEHOLDER ANALYSIS]. How does your framework advance [ETHICAL BUSINESS PRACTICE] while maintaining [THEOLOGICAL DISTINCTIVENESS]?

72. **International Ethics Construction**: Create sophisticated framework for

[INTERNATIONAL RELATIONS ISSUE] that integrates [THEOLOGICAL UNDERSTANDING], [INTERNATIONAL LAW], and [DIPLOMATIC THEORY]. How does your framework advance [INTERNATIONAL COOPERATION] and inform [THEOLOGICAL DISCOURSE]?

73. **Educational Ethics Development**: Develop comprehensive educational ethical framework that addresses [EDUCATIONAL CHALLENGE] through integration of [THEOLOGICAL ANTHROPOLOGY], [EDUCATIONAL THEORY], and [CULTURAL ANALYSIS]. How does your framework advance [EDUCATIONAL PRACTICE] while maintaining [THEOLOGICAL FOUNDATION]?

74. **Research Ethics Innovation**: Create original research ethical framework that addresses [RESEARCH ETHICS CHALLENGE] through integration of [THEOLOGICAL UNDERSTANDING], [RESEARCH METHODOLOGY], and [PROFESSIONAL ETHICS]. How does your framework advance [ETHICAL RESEARCH] while contributing to [THEOLOGICAL DISCOURSE]?

75. **Communication Ethics Construction**: Develop sophisticated communication ethical framework that addresses [MEDIA/COMMUNICATION CHALLENGE] through integration of [THEOLOGICAL TRADITION], [COMMUNICATION THEORY], and [CULTURAL ANALYSIS]. How does your framework advance [ETHICAL COMMUNICATION] and inform [THEOLOGICAL UNDERSTANDING]?

Doctoral Research Guidelines and Expectations

Research Categories for Doctoral Programs:

Doctor of Philosophy (Ph.D.) in Theology/Religious Studies:

- Emphasis on original theoretical contribution to theological scholarship
- Comprehensive engagement with international scholarly discourse
- Methodological innovation and interdisciplinary integration
- Preparation for academic career and scholarly publication

Doctor of Divinity (D.D.):

- Focus on constructive theological work for church and academy
- Integration of scholarly research with ecclesiastical leadership
- Contribution to denominational theological discourse
- Advanced pastoral and theological leadership preparation

Doctor of Ministry (D.Min.):

- Emphasis on practical theological research addressing ministry challenges

- Integration of academic research with ministerial experience
- Focus on improving ministry practice and church effectiveness
- Professional doctoral preparation for advanced ministry leadership

Substitution Guidelines for Doctoral Research:

[THEOLOGICAL TRADITION]: Specific denominational, confessional, or theological school traditions

[CONTEMPORARY CHALLENGE]: Current theological, cultural, scientific, or ministerial challenges requiring doctoral-level analysis

[METHODOLOGICAL APPROACH]: Specific research methodologies (historical-critical, phenomenological, ethnographic, comparative, etc.)

[INTERDISCIPLINARY RESOURCE]: Relevant non-theological academic disciplines and their methodologies

[PRIMARY SOURCES]: Original texts, documents, interviews, or data requiring doctoral-level analysis

[SCHOLARLY DEBATE]: Current academic controversies requiring doctoral-level contribution **[CULTURAL/CONTEXTUAL FACTOR]**: Specific geographical, ethnic, socioeconomic, or cultural contexts

Doctoral Research Standards:

1. **Original Contribution**: Each prompt should lead to original research contributing new knowledge to theological scholarship or ministry practice

2. **Comprehensive Literature Review**: Doctoral research requires exhaustive engagement with relevant primary and secondary sources

3. **Methodological Sophistication**: Advanced application of appropriate research methodologies with awareness of methodological limitations

4. **Interdisciplinary Integration**: Sophisticated engagement with relevant non-theological disciplines while maintaining theological integrity

5. **International Scope**: Awareness of and engagement with global theological discourse and scholarship

6. **Publication Quality**: Research should be suitable for peer-reviewed academic publication or professional application

Research Applications:

- **Dissertation Research**: Complete doctoral dissertation projects (200-400+ pages)
- **Academic Publications**: Scholarly articles for peer-reviewed journals
- **Conference Presentations**: Research presentations for academic societies
- **Grant Applications**: Research proposals for academic or ministry funding
- **Postdoctoral Research**: Advanced research projects for academic career development
- **Professional Consultation**: Expert analysis for churches, denominations, or organizations

Evaluation Criteria:

- **Scholarly Rigor**: Methodological sophistication and comprehensive research

- **Original Contribution**: New insights advancing theological knowledge or ministry practice

- **Theological Adequacy**: Faithfulness to Christian theological tradition while engaging contemporary challenges

- **Practical Relevance**: Connection between academic research and church/ministerial application

- **Global Awareness**: Engagement with international theological discourse and diverse cultural contexts
- **Professional Excellence**: Research quality suitable for academic publication and professional advancement

These doctoral-level prompts are designed to support the highest levels of theological scholarship and ministry preparation, requiring extensive research, original thinking, and significant contribution to theological knowledge and church practice.

REFERENCES

Aitken, R. L. (2019). Digital discipleship. Ministry Magazine. https://www.ministrymagazine.org/archive/2019/05/Digital-discipleship

Anthropic. (2025). Claude (Sonnet 4) [Large language model]. https://claude.ai Reason For Use: To organize prompts by category and classification.

Arega, T., et al. (2025). Constructivist instructional approaches: A systematic review of evaluation-based evidence for effectiveness. *Review of Education*, Early View. https://doi.org/10.1002/rev3.70040

Arrighi, N. (n.d.). ARRIGHI AI-C2 utilization spectrum: Five-stage learning spectrum for utilizing artificial intelligence (A.I.) in education [Educational framework]. Office of Teacher Education, Tennessee State University.

Atla. (2024, March 14). Why theological education should be on the frontier of artificial intelligence. Atla Blog. https://www.atla.com/blog/why-theological-education-should-be-on-the-frontier-of-artificial-intelligence/

Bond, M., Bedenlier, S., Marín, V. I., & Händel, M. (2023). Artificial intelligence in higher education: The state of the field. International Journal of Educational Technology in Higher Education, 20, 42. https://doi.org/10.1186/s41239-023-00392-8

Brown, T. K. (2023). Using digital discipleship to spread the mission and market your church. ResourceUMC. https://www.resourceumc.org/en/content/using-digital-discipleship-to-spread-the-mission-and-market-your-church

Chen, L., Chen, P., & Lin, Z. (2025). Design and assessment of AI-based learning tools in higher education: A systematic review. International Journal of Educational Technology in Higher Education, 22, 42. https://doi.org/10.1186/s41239-025-00540-2

Christian Theological Seminary. (2023, September 5). Faith, technology, and AI. https://www.cts.edu/2023/09/05/faith-technology-and-ai/

Churches, A. (2009). Bloom's digital taxonomy. *Educational Origami*, 4, 1-15.

Clinical Pastoral Education. (2025). CPE student guidelines and formation practices. Association for Clinical Pastoral Education. Retrieved from https://acpe.edu/education/cpe-students

Davis, F. D. (1989). Perceived usefulness, perceived ease of use, and user acceptance of information technology. *MIS Quarterly*, 13(3), 319-340.

Easy Cloud AI. (2025, June 6). Digital discipleship: Navigating AI's impact on modern mission work. https://easycloudai.com/2025/05/15/digital-discipleship-navigating-ais-impact-on-modern-mission-work/

García-Alberti, M., et al. (2024). Applying the technology acceptance model to online self-learning: A multigroup analysis. *Journal of Innovation & Knowledge*, 9(4), Article 100525.

Granić, A., & Marangunić, N. (2019). Technology acceptance model in educational context: A systematic literature review. *British Journal of Educational Technology*, 50(5), 2572-2593.

Hassell, R. (2024). ASCEND-AI: Elevating learning with smart AI prompts [Educational framework]. Smart Innovation Technology Center, Tennessee State University.

International Society for Technology in Education. (2017). ISTE standards for educators. International Society for Technology in Education. https://www.iste.org/standards/for-educators

Kawka, M. (2025). Artificial intelligence: A new challenge for human understanding, Christian education, and the pastoral activity of the churches. Religions, 16(8), 948. https://doi.org/10.3390/rel16080948

Klimova, B., & Pikhart, M. (2025). Exploring the effects of artificial intelligence on student and academic well-being in higher education: A mini-review. Frontiers in Psychology, 16, 1498132. https://doi.org/10.3389/fpsyg.2025.1498132

Klom, J., & Brown, V. (2024, October 16). Digital discipleship: Can churches grow people who may never attend in person? Outreach Magazine. https://outreachmagazine.com/features/82118-digital-discipleship.html

Liang, J., Stephens, J. M., & Brown, G. T. L. (2025). A systematic review of the early impact of artificial intelligence on higher education curriculum, instruction, and assessment. Frontiers in Education, 10, 1522841. https://doi.org/10.3389/feduc.2025.1522841

Liu, Y., et al. (2023). Extending technology acceptance model to higher-education students' use of digital academic reading tools on computers. *International Journal of Educational Technology in Higher Education*, 20, Article 32.

Melton, R. (2023). A.I. prompt rubric for education: AI-PromptScale rubric elements [Assessment rubric]. Tennessee State University Smart Center. https://ai-tnstatesmartcenter.org

Morrison, J. (2025). Opportunities and threats of artificial intelligence in Christian ministry: An interdisciplinary approach through the lens of scientific exploration and technology. Religions, 16(9), 1092. https://doi.org/10.3390/rel16091092

Nouwen, H. J. M. (1979). *The wounded healer: Ministry in contemporary society*. Image Books.

Ocen, S., Elasu, J., Aarakit, S. M., & Olupot, C. (2025). Artificial intelligence in higher education institutions: Review of innovations, opportunities and challenges. Frontiers in Education, 10, 1530247. https://doi.org/10.3389/feduc.2025.1530247

Oregon State University. (n.d.). Bloom's taxonomy revisited [Infographic]. Oregon State University Ecampus. Retrieved December 19, 2024, from https://ecampus.oregonstate.edu

Papakostas, C. (2025). Artificial intelligence in religious education: Ethical, pedagogical, and theological perspectives. Religions, 16(5), 563. https://doi.org/10.3390/rel16050563

Piaget, J. (1977). *The development of thought: Equilibration of cognitive structures*. Viking Press.

Richter, S., Giroux, M., Piven, I., Sima, H., & Dodd, P. (2025). A constructivist approach to integrating AI in marketing education: Bridging theory and practice. *Journal of Marketing Education*, 47(1), 32-48. https://doi.org/10.1177/02734753241288876

Schorr, A. (2023). The technology acceptance model (TAM) and its importance for digitalization research: A review. *Psychology Research*, 13(6), 245-267.

Steven, B. (2025, June 14). Digital discipleship. Bishop Steven's Blog. https://blogs.oxford.anglican.org/digital-discipleship/

Tennessee State University. (2025). *Policy on responsible use of artificial intelligence at TSU* (Policy No. 1.09). Office of Academic Affairs, Office of Technology Services, Office of Student Affairs.

Thompson, T. (2025, June). Theology and AI (Efficient or humanistic society). Theology AI Blog. https://www.theologyai.com/2025/06/

Tzeng, W. (2025). Leveraging artificial intelligence for theological education: 'AI is not a human, it is a tool.' Christian Daily

International.
https://www.christiandaily.com/news/leveraging-artificial-intelligence-for-theological-education-ai-is-not-a-human-it-is-a-tool

Vygotsky, L. S. (1978). *Mind in society: The development of higher psychological processes*. Harvard University Press.

Made in the USA
Monee, IL
21 April 2026

48713585R00155